'A brilliant summation of twenty years of "good farmer" scholarship, this volume retraces paths taken while pointing to terrain missed. The book offers a rich exploration into how moral identities are created and maintained among and between food producers. It is sure to become a key text among agrifood and rural studies scholars.'

— **Michael S. Carolan**, *Colorado State University, and Fulbright Distinguished Research Chair at University of Ottawa, Canada*

'This book not only invites us to go beyond the normative connotation of the "good farmer" but to engage in a theoretical informed, critical reflection on how the concept is being used by various groups. It is also a very timely reminder that to build sustainable food systems, we need to better understand the cultural dimension of agricultural practices.'

— **Ika Darnhofer**, *BOKU, Vienna, Austria*

'This outstanding collection brings together the leading scholars who have developed and applied the concept of the "good farmer". It provides a rich and timely summary of work to date and nuanced examples of exciting new thinking and future trajectories. This book will be a much-valued addition not only to the shelves of social scientists, but anyone looking for a richer understanding of the dynamics of agriculture and rural change.'

— **Mark Riley**, *Liverpool University, UK*

'For fifteen years, the idea of the "good farmer" has rapidly emerged as one of the most useful concepts in rural sociology and agrifood scholarship, yet it has never been subjected to a comprehensive theorisation and review. That moment has arrived. This book provides comprehensive and engaging insights by four of the leading scholars using this concept. The result is a benchmark work that will help establish the idea of the "good farmer" as a foundational concept in rural sociology, geography, anthropology and agrifood studies.'

— **Hugh Campbell**, *University of Otago, New Zealand*

THE GOOD FARMER

Developed by leading authors in the field, this book offers a cohesive and definitive theorisation of the concept of the 'good farmer', integrating historical analysis, critique of contemporary applications of good farming concepts, and new case studies, providing a springboard for future research.

The concept of the good farmer has emerged in recent years as part of a move away from attitude and economic-based understandings of farm decision-making towards a deeper understanding of culture and symbolism in agriculture. *The Good Farmer* shows why agricultural production is socially and culturally, as well as economically, important. It explores the history of the concept and its position in contemporary theory, as well as its use and meaning in a variety of different contexts, including landscape, environment, gender, society, and as a tool for resistance. By exploring the idea of the good farmer, it reveals the often-unforeseen assumptions implicit in food and agricultural policy that draw on culture, identity, and presumed notions of what is 'good'. The book concludes by considering the potential of the good farmer concept for addressing future, emerging issues in agriculture.

This book will be of interest to students and scholars of food and agriculture and rural development, as well as professionals and policymakers involved in the food and agricultural industry.

Rob J. F. Burton is a Principal Researcher at Ruralis – Institute for Rural and Regional Research, Trondheim, Norway.

Jérémie Forney is an Assistant Professor at the Anthropology Institute, University of Neuchâtel, Switzerland.

Paul Stock is an Associate Professor in the Department of Sociology and the Environmental Studies Program at the University of Kansas, Lawrence, Kansas, USA.

Lee-Ann Sutherland is a Research Leader in the Social, Economic and Geographical Sciences Department at the James Hutton Institute, Aberdeen, UK.

Other books in the Earthscan Food and Agriculture Series

A Global Corporate Trust for Agroecological Integrity
New Agriculture in a World of Legitimate Eco-states
John W. Head

Geographical Indication and Global Agri-Food
Development and Democratization
Edited by Alessandro Bonanno, Kae Sekine and Hart N. Feuer

Multifunctional Land Uses in Africa
Sustainable Food Security Solutions
Elisabeth Simelton and Madelene Ostwald

Food Security Policy, Evaluation and Impact Assessment
Edited by Sheryl L. Hendriks

Transforming Agriculture in Southern Africa
Constraints, Technologies, Policies and Processes
Edited by Richard A. Sikora, Eugene R. Terry, Paul L.G. Vlek and Joyce Chitja

Home Gardens for Improved Food Security
Edited by D. Hashini Gelhena Dissanayake and Karimbhai M. Maredia

The Good Farmer
Culture and Identity in Food and Agriculture
Rob J. F. Burton, Jérémie Forney, Paul Stock and Lee-Ann Sutherland

For more information about this series, please visit: www.routledge.com/books/series/ECEFA/

THE GOOD FARMER

Culture and Identity in Food and Agriculture

Rob J. F. Burton, Jérémie Forney, Paul Stock and Lee-Ann Sutherland

Routledge
Taylor & Francis Group
LONDON AND NEW YORK

earthscan
from Routledge

First published 2021
by Routledge
2 Park Square, Milton Park, Abingdon, Oxon OX14 4RN

and by Routledge
52 Vanderbilt Avenue, New York, NY 10017

Routledge is an imprint of the Taylor & Francis Group, an informa business

British Library Cataloguing-in-Publication Data
A catalogue record for this book is available from the British Library

Library of Congress Cataloging-in-Publication Data
Names: Burton, Rob, author.
Title: The good farmer : culture and identity in food and agriculture /
Rob J.F. Burton, Jérémie Forney, Paul Stock, and Lee-Ann Sutherland.
Description: Abingdon, Oxon ; New York, NY : Routledge, 2021. |
Series: Earthscan food and agriculture series | Includes bibliographical
references and index.
Identifiers: LCCN 2020015607 (print) | LCCN 2020015608 (ebook) |
ISBN 9781138727793 (hardback) | ISBN 9781138727960 (paperback) |
ISBN 9781315190655 (ebook)
Subjects: LCSH: Farmers—Psychology. | Agriculture—Social aspects. |
Agriculture—Psychological aspects.
Classification: LCC S494.5.P76 B87 2021 (print) | LCC S494.5.P76
(ebook) | DDC 630—dc23
LC record available at https://lccn.loc.gov/2020015607
LC ebook record available at https://lccn.loc.gov/2020015608

ISBN: 978-1-138-72779-3 (hbk)
ISBN: 978-1-138-72796-0 (pbk)
ISBN: 978-1-315-19065-5 (ebk)

Typeset in Bembo
by Apex CoVantage, LLC

CONTENTS

PREFACE

In a world where technology is playing an ever-increasing role in food production, farms are becoming ever larger, and commercial imperatives are infiltrating all aspects of agriculture, the reader may rightfully question the relevance of a book wrestling with the concept of the 'good farmer'. Does culture and identity in agriculture really matter? Here, we argue that it matters now more than ever. Agriculture has constantly evolved since the first human settlement, but it is not the technologies, crop varieties, or management practices that drive production; it is the people who manage the land themselves. On one hand, each farmer makes individual decisions on how to farm his/her land. On the other hand, each farmer is a member of a community of farmers facing similar choices and challenges – one that is tied together through a key common purpose, to produce food. In this community, being a 'good farmer' is about more than having the highest income; it is about working together, building self-esteem, enjoying the lifestyle, and, from time to time, being recognised by others. The notion of 'good farming' is the glue that binds farmers together, provides regional distinctiveness, maintains cultural landscapes, upholds rural communities, and motivates farmers and families to keep farming when commodity prices suggest a more economically rational approach would be to give up. This is what keeps food on our tables and maintains our countryside and its communities. We offer this book as a testament to the farmers we have known, the stories they have told us, and in the hope that there will be generations of farmers to come to argue about whether someone is a 'good farmer' or not.

A note on how we wrote this book

This book is a monograph and not an edited volume, with authors ordered alphabetically rather than necessarily reflecting the extent of their contribution to the book. With the exception of Chapter 3 on the historical development of the symbolic meaning of 'tidy farming', the first drafts of the chapters were written in teams of two (using various author combinations), and then, following discussion within the writing team, opened for further additions or changes based on collective agreement. There were, of course, disagreements between the authors – and some of these still stand. However, the content of the book was decided through consensus with any remaining disagreement existing only to the extent that an individual author might possess a range of thoughts on an issue. The advantage of writing in this manner, difficult though it sometimes was, is that we bring together expertise from across a range of disciplines and are able to address a concept to which no single discipline has a unique claim. We look forward to continuing this conversation, not only with farmers, but policy makers, fellow academics, and anyone who cares about agriculture, food, and farmers.

R.J.F.B
Trondheim, Norway

J.F.
Neuchatel, Switzerland

P.V.S
Dunedin, Aotearoa New Zealand

L-A.S.
Aberdeen, Scotland
March 2020

ACKNOWLEDGEMENTS

Rob would like to thank Jérémie, Lee-Ann, and Paul for the many inspiring conversations and sheer hard work that went into completing this book. It was a roller coaster ride, but we made it. I am also grateful to the Norwegian Research Council for continuing to fund my research and to Ruralis where being surrounded by great colleagues makes working both productive and enjoyable. Finally, many thanks to Heike Fischer for being my inspiration and, often, my academic adviser, and to Marita and Quinn for being the best kids in the world.

Jérémie would like to thank Rob, Paul, and Lee-Ann for engaging together in this intellectual journey. Also thanks to Yvan Droz and Valérie Miéville-Ott who guided him in his first explorations of farming cultures in Switzerland in the early 2000s. Special thanks to Virginie, Lili, Joé, Mina and Tim.

Paul would like to thank Rob, Jérémie, and Lee-Ann for an enjoyable and thought-provoking writing journey. This book has been long in the making. Thanks also to the Centre for Sustainability at the University of Otago in Dunedin, NZ and especially Hugh Campbell for fostering many of the conversations reflected in this book and the collaborative spirit on display. Thanks also to Colleen, Liam, Audrey, Emery, and Maisie.

Lee-Ann would like to thank Rob, Jérémie, and Paul for their enthusiasm and persistence throughout this process. My writing time was made possible by the James Hutton Institute and Scottish Government's Rural Affairs, Food and the Environment Strategic Research Programme (2016–2021). Thank you to my parents, who gifted me with a love of farming and are still demonstrating what it means to be 'good farmers' in their 70s. Appreciation and much love to Brian, for keeping the home fires burning so I can do this job I enjoy so much.

The authors would like to recognise and acknowledge the contribution of Alex Burdge and Ruth Stamper, who worked as research assistants that contributed especially to chapters 2 and 5. They were supported through the Kansas General Research Fund allocation # 2172081.

We also acknowledge the many farmers who have spoken with us over the years over cups of tea, or at the breakfast table after the morning milking, and who have willingly shared their memories and everyday lives. We recognise that farming is an honourable and, at times, difficult pursuit. We offer this book as a thank you, hoping it will contribute to a better recognition of their perspectives and help all of us to sometimes look through the good farmer's eye.

1

THE 'GOOD FARMER'

Cultural dimensions of farming and social change

Throughout the world, farmers gather in groups large and small to discuss the weather, crop prices, the government, and a host of other factors affecting their farms and daily lives. Around Formica or oak tables, leaning against a piece of old farm machinery, or wherever they happen to be, these same farmers share anecdotes, jokes, and apocryphal stories about what happened to the farmer down the road, whose crops are full of weeds, whose tramlines are crooked, who sold his/her sheep for the highest price, who bought a new tractor, and so on. Within these settings the topic invariably turns to who is a 'good farmer' – and who is not. In the daily lives of farmers, the act of calling someone a 'good farmer' is an acknowledgement of appreciation and status within the community. It is not a title given lightly. Rather it bestows on the individual recognition of the cultural competences that make a farmer worthy of being sought out by others for assistance – knowledge, skills, or material assistance – the provision of which, in turn, forms bonds of mutual obligation within farming communities. It is the centre of the farming culture. The concept of being a 'good farmer' is global, with a 'good farmer' being recognisable across different regions and over generations – providing the understandings of what it is to be a 'good farmer' are shared.

While rural social scientists and extension agents throughout the world speak in the language of this everyday usage of what and who is a good farmer, it is only within the past 15 years that social scientists have taken the concept of the 'good farmer' seriously as a category important in and of itself. Rob Burton's article "Seeing Through the 'Good Farmers' Eyes: Towards Developing an Understanding of the Social Symbolic Value of 'Productivist' Behaviour", published in *Sociologia Ruralis* in 2004, helped establish an intriguing and inviting space for reinvigorating a lapsed engagement for agri-food scholars. By following the lead of the farmers, rural social scientists have hit upon a cultural and political category important not only to the farmers themselves, but to extension agents, policy makers, environmental and development NGOs, non-farming neighbours, and increasingly, farmers with multiple, often contested perspectives on who is a good farmer. The 'good farming' literature has grown steadily in the past 15 years, signifying important changes both to the rural social sciences and agri-food studies as well as to the worlds of farming.

At the time Burton's (2004) paper was published, agri-food studies and rural studies of farmers' decision-making strategies were still strongly influenced by rational models of decision making, although a gradual move away from these models had been occurring. In particular, Wolpert (1964) had argued that farmers did not make decisions on the basis of economic rationality alone. They were not interested in maximising outcomes, only reaching a satisfactory solution. Later, Gasson (1973) contended that economically optimal outcomes were not reached because farmers' decisions were not always driven by economic goals, but that sometimes farmers sought to optimise social, intrinsic, and/or expressive goals. The development of a reliably predictive attitude-behaviour model by Fishbein and Ajzen (1975) saw the interest turn to attitudes to explain non-economically rational behaviour – enabling agricultural studies to identify what it was farmers were prioritising (e.g. Ilbery, 1983) – and, by the mid-1990s, the "behavioural approach" was being widely employed. This was outlined in Morris and Potter (1995) as being based

> on the motives, values and attitudes that determine the decision-making processes of individual farmers.
>
> *(p. 55)*

Of course, farmers were not able to simply do whatever they wanted. Rather, the decisions made were governed by structural factors that were seen to constrain, facilitate, and reflect the motivational preferences of the farmer or farm family.

Beginning in the 1990s agricultural studies pushed beyond the economic/attitudinal approaches and began looking at the notions of identity and culture as a potential factor in understanding farmer behaviour. To some extent this represented the continuation of a tradition of exploring role and identity emanating from gender studies – for example, the role of women as business managers (Hastings, 1988), the subordinate nature of women's roles in agriculture (Stebbing, 1984; Gasson and Winter, 1992), the gendered division of labour roles and decision making (Sawer, 1973; Berlan Darqué, 1988; Whatmore, 1991), the role of drinking in establishing masculine rural identities (Campbell, 2000), and the construction of masculine farmer identities (Saugeres, 2002). However, despite the realisation of its importance to gendered behaviour in agriculture as early as the 1970s it was not until the turn of the century that researchers thought of applying these same concepts to what were still generally perceived as the 'rational', 'economically oriented' male decision makers managing farms.

Some mention had been made of the 'good farmer' in the 1990s. Fairweather and Keating (1994), for example, briefly observed the social importance of being recognised as a good farmer. Phillips and Gray (1995) in an Australian study examining sustainability in agriculture, observed that there were a number of competing definitions of what it meant to be a 'good farmer' within the farming community. They offer one of the earliest recognitions of the social importance of being a 'good farmer' as they note:

> farmers are constantly engaged in strategic farm practice to gain the community recognition required to define their farming practice as 'good farming'. The social relations of farming are thus both enabling and constraining to farming practice.
>
> *(p. 131)*

Harrison et al. (1998) also noted how recognition as a 'good farmer' was important and identified distinctions between what farmers and non-farmers viewed as a 'good famer'. Throughout the decade references to the 'good farmer' increasingly came to acknowledge the term as a cultural construct. Perhaps the first paper to outline this as part of a cohesive theory was Tsouvalis et al. (2000). In seeking to understand moral conflict between different knowledge-cultures (i.e. "socially negotiated structures of meaning that enable and constrain social actions", p. 912), Tsouvalis and colleagues observed that the boundaries between cultures were determined by notions of what it meant to be a 'good farmer'. Knowledges that fell outside cultural definitions of the 'good farmer' were contested on that basis, e.g.,

> 'good' farmers are supposed to know their land; 'the public' and 'politicians' don't understand 'us'; 'scientists' and 'manufacturers' should come and talk to 'us'; 'we' know the land, 'we've' farmed for years.
>
> *(p. 922)*

While Tsouvalis' work addressed the introduction of precision farming, we argue it was the introduction of multifunctionality objectives to the Common Agricultural Policy in the 1980s and 1990s that provided the impetus for a new focus on culture and identity in farmer studies. Prior to this, farmers had been entreated to produce as much as possible as a post-war civic duty. However, with oversupply of food products and environmental degradation, policy makers – in Europe in particular – began an attempt to coax farmers away from ever-increasing production through the introduction of agri-environmental and diversification schemes.

Although some farmers took to these changes willingly, for others the resistance was considerable. New roles as conservation managers and providers of services to the public led many farmers to question the very essence of who they were. In the case of attempts to promote tree planting, in particular, the concern was frequently expressed that they were "farmers, not foresters" and "farmers farm" (Williams et al., 1994; Allison, 1996; Burton, 1998). Lowe and colleagues noted as early as 1999 that participation in agri-environmental schemes for a decade should have led to "discernible changes" in farming attitudes and culture (Lowe et al., 1999), but over a decade later there was still little evidence of a broad cultural shift. It was an attempt to understand the role of identity resistance to adopting the forestry and recreation provision roles promoted by the UK Community Forest scheme at Marston Vale, Bedfordshire, that led Burton (1998) to look at the importance of identity to farmer decision making and through this work the social symbolic importance of farming behaviours was first described in detail.

The initial study was an application to 60 farmers of a quantitative survey based on the identity theory of Sheldon Stryker (1994) – itself a 'social structuralist' interpretation of Mead's (1934) symbolic interactionist perspective (see Chapter 4). During the conducting of the survey the theme of the 'good farmer' emerged repeatedly as farmers sought to explain why it was that forestry was not farming. This led to follow-up series of 13 qualitative re-interviews conducted to specifically address the issue of the 'good farmer' by asking questions such as: When does a farmer adopting new roles cease to be a farmer? When can a newcomer to farming be considered a proper farmer? What symbolises belonging to the farming community? Can a woodland be viewed as a crop? How are symbols of good farming transferred along the roadsides? (see Burton, 1998). The publishing of the core of this PhD thesis in 2004 in the "good farmer" paper was at the forefront of a wider cultural turn that occurred in both English (e.g. Cloke, 1997) and non-English (e.g. Billaud, 1991; Droz and Miéville-Ott, 2001; Mallein and Cautrès, 1993; Sylvestre, 2002) rural studies in the 1980s and 1990s – and thus was well-positioned to become an important text.

Figure 1.1 shows the citations of the paper year on year up to the end of 2019 and illustrates how the paper has gradually been established as a seminal text

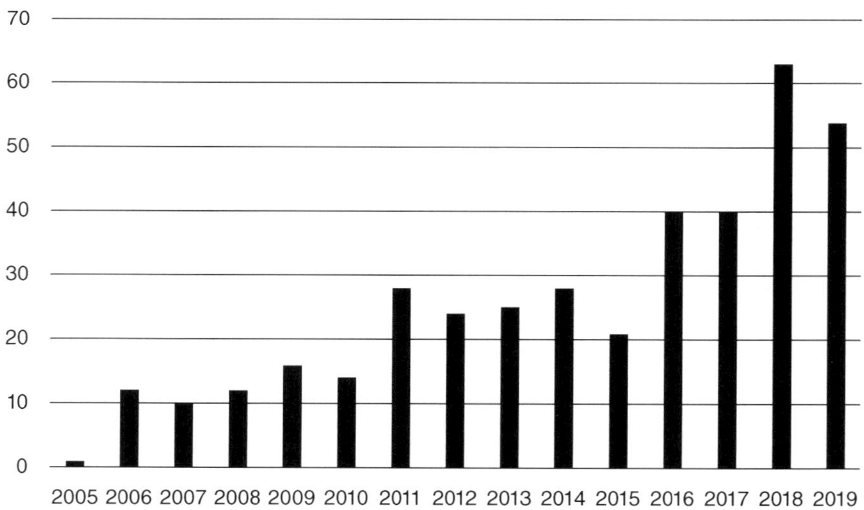

FIGURE 1.1 Citations of the 'good farmer' paper (Burton, 2004) by year according to Scopus

amongst international researchers, with over 60 citations of the work in 2018 alone – 14 years after the publishing of the original paper.

We argue that this growth in usage reflects the quality of the concept and its increasing relevance for the challenges faced by today's farmers. It is interesting to observe where the 'good farmer' has been applied. For the most part citations appear in social science publications related to rural and agricultural issues, often addressing the complex relations between farmers and environmental management. These papers refer to the paper in a very general and succinct way – citing it in order to mention the identity dimension of farming rather than engaging with development of the concept itself. Further, they often highlight only one aspect of the 'good farmer' approach. For example, some point to the social pressures exerted on individual behaviours (Chan et al., 2017; Isgren, 2016) while others mention the tension with government action and policies (Caputo and Butler, 2017; Collins et al., 2016; Hansson et al., 2012).

A further group of papers reference Burton (2004) to confirm presumptions about farmers' identities or relations to the environment. These authors again do not engage with wider theorisation of farming cultures, but rather refer (often only briefly) to the paper as a proof of the existence of a farming culture. Sometimes, these patterns are applied to farmers as a very general and uniform category across time and space, generations, and geographies. Burton (2004) also represents a significant reference illustrating the strength of productionist behaviour in Western farming (how it has become part of the culture – McGranahan, 2014; Van der Wal et al., 2014), its relation to the valuation of tidy landscapes and crops (Cardoso et al., 2016; Pouta et al., 2014; Tautges et al., 2016), and farm succession (Brown, et al., 2014). While the paper certainly illustrates the importance of production behaviours for defining the farmer identity, production roles were the only aspect of farming the paper addressed – thus it should not be used as an illustration that production is the only thing that matters to farmers.

In addition to the social science papers, we found a significant number of references by researchers from other disciplines (around 44 papers in Scopus prior to late-2019), generally again related to environmental issues in farming. A first group of papers engage seriously with the 'good farmer' approach and seem to integrate part of the more theoretical developments and sometimes later publications, although never in great depth. This is understandable as they are not primarily addressing social science issues. The 'good farmer' approach appears here as a way to question rational choice theorisations of farmers' behaviours (Jakobsen, 2017; Kristensen and Jakobsen, 2011a, 2011b), to include in the analysis the impact of the social and cultural context (Lankester, 2013; White and Selfa, 2013; Barnes et al., 2015), and to demonstrate the concept of social control, notably through the idea of "roadside farming" (Sandberg and Jakobsson, 2018; Seabrook et al., 2008).

What drew us to writing this book is that, while there is now an abundance of literature looking at or using the concept of the good farmer, there has been little attempt to develop some of the key concepts. Two omissions stand out. First, most of Burton's work has focused on understanding the symbolic importance of the practice of farming (in particular, raising crops) – rather than other non-production roles associated with being seen as a good farmer. As a result, the concept as Burton outlines it is restricted to a relatively small (albeit significant) aspect of the construction of 'good farmer' identities. Second, neither Burton (2004) nor any of his later publications attempt to explain *how* behaviours in agriculture become more or less symbolically important. While it is implicit in the 'good farmer' concept that the symbolic value of behaviour *must* change as technologies, management approaches, crops, and so on, change, the issue of how and why this occurs is not outlined in Burton's paper. This left a significant hole in the theoretical construction of the good farmer, but, despite the existence of a few papers that have dealt with changing constructions of the 'good farmer' (e.g. Sutherland and Darnhofer, 2012), the issue of symbolic change has yet to be adequately addressed.

As a result of the broad range of uses and users of the 'good farmer' concept the literature is rather dispersed. One of our objectives here is to draw this literature together into a single volume for researchers and all those interested in the 'good farmer' concept, as well as to address some of the significant gaps in the concept. The increasing use of the seminal paper on the 'good farmer' shown in Figure 1.1 reflects how the 'good farmer' is increasing in relevance due to increasing urgency in addressing agriculture's role in climate change (e.g. through 'climate smart agriculture'), continued biodiversity loss in Europe, the loss of traditional ecological knowledge, the increasing mechanisation of farming roles, and the return of food security to the political agenda, prompting reflection on whether farmers are heading towards 'neoproductivism' (Wilson and Burton, 2015). In the remainder of this chapter we review the use of the concept in the literature, observe the value of the 'good farmer' approach as more than a form of 'resistance' to new policy approaches, and outline the content of the book. Before reading on, we ask that scientists of any discipline referring to the 'good farmer' as an explanatory concept should do it with a strong theoretical understanding while recognising both that there is (as yet) neither a 'theory of the good farmer' nor a single definition of what a 'good farmer' is or is not.

Multiple uses of the 'good farmer' term

With its use in a broad range of disciplines as well as academic and non-academic usages, the term 'good farmer' has been employed in multiple – often somewhat overlapping – ways. It is difficult to discern distinct categories of use for the term, but they broadly fall into those employed by farmers themselves, those employed by agricultural/natural scientists and policy makers seeking to change agricultural practices, and those employed by social scientists in order understand cultural influences on farmer behaviour.

Farmer-based uses

1 A commonsense category used by farmers to refer to a farmer whose farming behaviours reach a certain level of competence.

2 An emic category used by farmers to describe the social significance of other farmers' work. Emic perspectives focus on how intrinsic cultural distinctions are meaningful to members of a social group. The concept of the 'good farmer' has emic origins in the academic literature as a reporting of farmer's own accounts of the symbolic significance of productivist behaviour – with a focus on shared symbolic understanding and the transfer of information on who is a 'good farmer' between members of the community.

Agricultural/natural science and policy-based uses

3 A standard assessed by agricultural/natural scientists or industry members about the productivity and/or environmental impact of a particular farmer's practices (i.e. who is objectively a 'good farmer' on the basis of the fit between their farming practices and expert scientific knowledge or industry/policy goals).

4 A pre-scientific agriculture (mainly 18th century and early 19th century) means of asserting the validity of one's own experimental practices without presenting evidence. In this case the term referred to anecdotal visits by generally unnamed 'good farmers' from the neighbourhood who, invariably, proclaimed the success of the approach.

5 A rhetorical means of laying claim to a particular approach to agriculture as the 'right' approach by those seeking change. This use of the 'good farmer' was employed by the agricultural improvers of the 18th and 19th centuries to aid their attempts to move agriculture away from traditional towards scientific practices. More recently the phrase 'good farmer' has arguably been replaced by other terms such as 'entrepreneurial', 'innovative' or 'sustainable' in accordance to scientific knowledge, or the political definition of a desirable agricultural future.

Social science based uses

6 An academic conceptualisation of the symbolic capitals and cultural scripts that are generated or promoted in farming practices, most often in relation to understanding farmers' apparent resistance to attempts by policy makers to introduce change through voluntary payment mechanisms. Here the 'good farmer' concept (via e.g. Goffman or Bourdieu) offers a theoretically supported means of explaining the lag effect in policy adoption as a form of cultural preference.

7 A commonplace reference amongst social scientists for social norms in agriculture. While this can reference the theoretically informed literature mentioned

previously, this is not necessarily the case (i.e. it can be used in a common sense manner) and where the theoretical literature is referred to it is not developed.

8 A methodological tool used by social scientists in research to elicit responses relating to social norms (e.g. what would a 'good farmer' do?). Asking a study participant to identify and describe a 'good farmer' in his/her locale provides a useful entry into descriptions of normative expectations, in way that is comfortable for farmers – they can relate to the idea of 'good farming' quite naturally, whereas asking specifically about norms or social expectations would be more challenging.

9 A moral/ethical standard where two or more concepts of who is a 'good farmer' are in conflict with one another – derived through 'grounded' academic analysis of empirical research. This is particularly evident when multiple moral positions are possible. For example, the responsibility to future generations of addressing climate change through reducing the intensity of agriculture can be countered by arguments that producing as much as possible is necessary in order to feed the world today. Both arguments involve a relatively high level of abstraction and have valid moral foundations.

It is notable that whether used by farmers, natural/agricultural scientists, or social scientists, the essential meaning remains the same – the 'good farmer' term is used to refer to the extent of cultural competency/morality in farming either by those within the peer group, those seeking to influence the peer group, or those studying the cultural construction and functioning of the peer group. Attaching the term 'good farmer' to specific practices within agriculture (both implicitly and explicitly) enables the transfer of an evaluation of the practices along this axis of cultural competency/morality from good to bad. The term 'good farmer' rather than 'good farming' (commonly used to refer to good farm management) is used as it refers to both the production roles and the wider social and moral roles of the good farmer. Participation in local community events or raising children to take over the farm may not be described as 'good farming' but can be an integral part of the 'good farmer'. Further, good farming may refer to practices performed in a single year – whereas it can take a lifetime to establish (and a lifetime to disestablish) oneself as a 'good farmer' as each year offers new challenges that must be met with a similar level of proficiency. Any farmer can invest in the best new machinery and buildings and farm at a loss, but it is maintaining the family name and reputation on the land over generations that demonstrates one is a 'good farmer'.

The 'good farmer' as a form of resistance

As noted previously, the term 'good farmer' was used historically as a means for those concerned with the future of agriculture to lay claim to a particular form of farming as virtuous while, at the same time, denigrating customary farming methods. Whether in the form of agricultural improvers seeking to professionalise and industrialise agriculture in the 18th and 19th centuries, Seaman

Knapp and the demonstration farm movement in early 20th century America, or attempts to create Community Forests through voluntary tree planting in late 20th century England, farmers have always been faced with competing visions of what it is to be a good farmer. This has helped agriculture serve the changing needs of society over the centuries. However, these changes often led to conflict with established notions of the 'good farmer' as symbols of good farming associated with the new forms of agriculture arose to challenge the old (see Chapter 3). Since World War II European agriculture (as well as other agricultures to a lesser extent and/or during particular periods – even neoliberal New Zealand) has been heavily subsidised which has allowed the state to play a critical role in determining what a 'good farmer' is and is not. 'Good farmers' as determined by the state are financially rewarded, whereas 'bad farmers' are not. As notions and symbols of the 'good farmer' are deeply embedded within farming cultures, these new definitions of good and bad farming have not always been immediately accepted by farmers and, where this occurs, the embedded notion of the 'good farmer' is often employed as a form of resistance to role changes.

It is important to clarify that resistance need not only refer to activist social movements or political lobbying. In many respects, farmers' resistance happens at the level of social and moral norms and everyday practices, for example, to modernist conceptions of development that perceive progress as equivalent to the extinction of peasants and farmers. The idea that farmers (or some of them) resist change is evident in the normative uses of the 'good farmer' (or other terms referring to the cultural dimension of farming) as a common sense category which promotes and defends a specific interpretation of farming practices and denigrates others. It is also often present in analytical approaches, where the cultural dimension is generally used to explain why farmers do not behave as other – often more powerful – actors want them to behave.

While, in many cases, the cultural dimension is used as an analytical and neutral concept, it can also incorporate a moral inclination in its implicit relation to change. This moral tone can follow two different inclinations. On the one hand, it endorses the acceptance of the fundamental good in the change that is promoted (e.g. environmental farming practices supported by the state) and the cultural dimension of farming is used to explain and understand the mechanisms producing resistance among farming communities. One of the implicit and underlying foci is therefore on how to facilitate the acceptance of change, and, in the longer term, to provoke adaptation of the cultural definition of farming. On the other hand, a more critical approach describes farmers' cultural resistance as a questioning of the 'good' lying within the change promoted by governing agents, and a fight against the power of hegemonic forces. Arguably, this perspective tends to side with farmers in contesting the value of change and highlighting its negative social or environmental impacts.

What characterises the scholarship on the cultural dimension of farming is this double focus on the articulation of change and power in a general movement that produces a new moral definition of good in farming and aims to make farming and farmers 'better', according to this new moral order. This "will to improve" (Li, 2007) is typical of a modernist and developmentalist approach, where elites (urban, white, educated) and institutions of power (state, industries, agricultural schools) consider it their right and duty to change and correct lower classes' habits and cultures. The origin of the 'good farmer' approach thus lies in the question of changing farming cultures. However, the idea of change is not neutral, nor obvious: it implies (1) an underlying cultural and moral definition of the good in farming and (2) a set of power relations that allows the imposition of this definition as 'good' for farming.

In taking this perspective, there is a high risk of forgetting that change is constant. The backwards or conservative farmer or peasant remains a stereotype for many agronomists and developmentalists. The logical conclusion would be then that if change is needed it has to be promoted by pushing or forcing farmers to change, through external interventions: regulations, financial incentives, education and knowledge transfer, etc. This perspective is impregnated by an understanding of top-down systems of knowledge, where experts provide evidence to decision makers who have to implement this knowledge by convincing practitioners to obey. In such a framing there is little room for freedom or reflexivity in given contexts about the right to govern one's own land.

The 'good farmer' approach, when applied from a sound theoretical and conceptual basis, critiques classical interpretations of farmers' resistance to change. As expressed by Haggerty et al. (2009), the theorisations of the 'good farmer' allow us "to move beyond simple narratives that locate farmer reluctance to change within tropes of 'tradition' or agrarian conservatism" (p. 769) by showing how they are involved in struggles for identification and social competition. Highlighting the role of farming cultures is important if we are to understand the rationality of farmers' behaviours. Within the environmental management literature, it is increasingly accepted that changing farmers' attitudes to conservation behaviour provides temporary change (often linked to payments) whereas cultural change – changing socially embedded understandings and symbols of good farming – provides a much more long-lasting solution (Burton and Pargahawewa, 2011).

However, there is always a risk of focusing on the cultural resistance without questioning the rationality and good of the governance action, or other factors related to the wider agri-food system. A shallow understanding of the 'good farmer' framing can lead to the assumption that the problem lies in farmers' minds. A deeper sociological theorisation, to the contrary, contradicts such shortcomings. First, it points to the relational and systemic nature of the resistance to change, by showing that the cultural definition of farming reflects a history of power relations, where a multitude of actors intervene and interact – not only the farmers. In other words, the 'good farmer' framing helps to demonstrate how farmers' resistance to change reflects, in fact, the resistance to change of the whole agri-food system. Therefore, policies or others' actions that aim to change farming cultures without impacting on how farmers are positioned within the system are likely to fail. As we will indirectly show in this book, cultural changes never happen alone. While understanding the cultural dimension of farming is necessary for transforming agri-food systems, to focus on farmers' cultural resistance alone is a dead end.

Moreover, 'cultural resistance' to change is not a bad thing per se. Put bluntly, it is useful to remember that cultural definitions of farming exist in part because they provide solutions to social needs. In Chapter 4, for example, we describe how farmers in 18th century Buckinghamshire (UK) resisted the levelling of anthills on meadows for over 100 years despite pressure from landlords to "improve" fields because the reeds that grew in the hollows between the anthills were needed to provide straw. Cultural definitions set moral limitations on individual behaviours through social control (e.g. limiting greed, unfairness, and/or amoral behaviours). They constitute collective forms of knowledge and cultural capital that strengthen the resilience of communities. They also create a buffer against rapid changes that might also be misleading or erroneous in the longer term. Cultural resistance also allows a cautious and gradual integration of new practices in the community. In addition, the stable nature of farming communities – often related to the inter-generational dimension of farms – means that networks of social relationships become strong and complex (intertwined) because they are based on mutual obligations developed through exchange of cultural (and possibly economic) capital. The maintenance of these networks has an important role in agriculture. This is particularly the case in the current environment where economic returns are often low relative to those of non-agricultural occupations. In this framing, 'resistance' is evident when farmers act – sometimes subconsciously – to protect developed capital systems by seeking to continue the production of existing symbols of 'good farming' in the face of externally imposed threats.

The 'good farmer' can serve also as a mirror reflecting the imperfections of the transformations promoted by powerful agents, such as state agencies, international organisations, influential NGOs, or dominant corporations. As demonstrated in the literature (e.g. Forney, 2016; Haggerty et al., 2009), when farmers resist the logics of specific agri-environmental schemes, often they do not criticise the fundamental necessity of caring for the environment. Rather, they point to imperfections in scheme design such as the loss of autonomy resulting from structural dependency on public money or the excesses of bureaucratisation. Often farmers claim to be the 'true environmentalists', because it is 'not in their interest' to allow the environment to degrade and, somewhat erroneously, that the fact there is something left to protect after hundreds of years of agriculture indicates that farming is intrinsically environmentalist. Exploring these resistances is useful not only to find ways of overcoming them, but also to reflect critically on the practices of governance developed by actors that have the power to influence the future of food systems.

This combination of perspectives implies a move away from an implicit understanding of cultural values, as opposed to rational thinking, and as an obstacle to change that is still largely present and reflects a modernist and developmentalist paradigm. We contend that this reflects a misunderstanding of what culture is. One of the authors of this book worked for a short period within an agricultural research organisation that maintained significant industry connections, where success in agriculture was viewed as a matter of survival of the fittest (most productive). In this environment, culture was something only indigenous populations possessed and was customary, primitive, and non-scientific. Rather than part of farming, culture was seen as antagonistic to good, efficient, modern, agriculture. An alternative conception of the 'good farmer' acknowledges the inescapable presence and influence of the cultural dimension and aims to integrate it within the construction of better food futures. This does not mean that current definitions of the 'good farmer' will not evolve. On the contrary, they have to and do so constantly (see Chapter 3). As recalled by Emery (2010), culture is used and manipulated, and continuously re-made and re-interpreted. But culture also rests on elements of continuity, of a sense of belonging to something that goes beyond the self and connects to the past and future. This means that current cultural definitions of the 'good farmer' serve as a basis for developing new interpretations, hopefully interpretations that are better adjusted to the challenges that agri-food systems will face in the future through the combination of old and new values.

What you will and will not find in this book

In a book about the 'good farmer' readers might reasonably expect to find an answer to the question, Who is a good farmer? The premise that there is a single 'good farmer' who can be defined and replicated in order to optimise the environmental, economic, and productive capacity of the food system is one that doubtless appeals strongly to policy makers seeking to build a resilient food system. However, we contend that there are two reasons this is not the case. For one, uniformity could come at a cost. A universal 'good farmer' focused on symbols of environmental management might not produce sufficient food, a universal production oriented 'good farmer' might lead to the loss of biodiversity, a universal 'good farmer' who always economically optimises could cause dramatic swings in commodity prices, while a universal 'good farmer' focused on community development and maintaining younger people in agriculture alone could struggle to make agriculture profitable. It is diversity in agriculture that creates a resilient food system, not uniformity of farmer types. Second, as any farmer will tell you, each farm is different and faces a unique set of challenges due to factors such as topography, farm size, location, and so on. Although some of the roles (such as caring for the land and passing the farm on to the next generation) and some symbols (such as tidy farming) of the 'good farmer' are more or less universal, each farmer needs to adapt to the specific context within which the farm is located such that a 'good farmer' in one region may not be a 'good farmer' in another. Instead of building a picture of a universal good farmer, this book should leave the reader with an impression of how contested the definition of a 'good farmer' is.

There are two key features we would like readers to be aware of when reading this book:

First, the book does not advocate for any specific form, style, or type of farming practice. While there are significant challenges facing global agriculture that deserve policy attention, our answer as social scientists is to defend the idea that discourses about good and bad farmers have to be considered not only as normative prescriptions, but also as a reflection of the cultural and identity dimension related to practices of agriculture. If we, as societies, want to find effective solutions to social challenges we need to take this dimension seriously, as a basis for moving towards more hopeful food futures. Rather than identifying a single 'good farmer', here we provide both historical and thematic overviews of the emergence of a significant umbrella concept – the 'good farmer' – that bears academic scrutiny by virtue of the important role it plays in maintaining cultural farming traditions and their evolution over time in specific contexts. In the end we do not proclaim a 'good farmer' the winner – as in who or what is the best kind of farming – but rather we offer a pathway to understanding how people come to make those kinds of declarations in specific times and places and what consequences there are for the decision makers, farmers, animals, land, watersheds, and places.

Second, we do not present a singular 'theory of the good farmer'. As authors we have each conceptualised the 'good farmer' differently – as have the many other authors who have employed the concept. Rather, we propose that conceptualisations should be selected on the basis of their utility for understanding the particular aspect of the cultural object/behaviour under investigation (see Chapter 4). While we have all been involved in the development of the current conceptualisation of the good farmer, we come from different academic backgrounds – environmental sociology, rural geography, and the anthropology of food and agriculture – and, as is fitting for a widely applied concept such as the 'good farmer', we hold different perspectives on how the 'good farmer' concept is best operationalised. In reading the book, it is thus possible, in places, to notice changes in the style of writing that reflect different disciplinary approaches. We have aimed to make the book consistent, but as the content of the book represents the work and views of all four authors equally, there is a degree of variation within the book that may not be expected from a single authored monograph. We see this as a strength of the book, rather than a weakness.

Book and chapter overview

The question of what to include in a book on the 'good farmer' is a vexing one. Almost all issues relating to agriculture that have been subject to research in recent years are of relevance to the concept of the 'good farmer'. In addition, many different theories have been used in conjunction with the 'good farmer' concept (see Chapter 4), and there is consequently much that could be discussed with regards to theoretical conceptualisation. In making our plan for the chapter content we decided to address issues where there is either a significant gap in the literature or a need to draw together the existing literature. This resulted in a book of eight chapters with the first, naturally enough, introducing the topic. The remaining seven chapters are as follows:

In Chapter 2, "The origins of the 'good farmer'", we present an original analysis of the historical emergence of the 'good farmer' in the literature on farming and agriculture. Exploring both the British (1500–1800) and the US (1700–1945) literatures, the chapter illustrates how normative discourses on what a good farmer (and/or husband) should be and do have evolved over time. This historical journey details how the 'good farmer' (or good husbandman) as a family, community and godly man became the 'good farmer' as a production-maximising food provider. The chapter explains how the view of the 'good farmer' as a 'productivist' is a more recent construction, and consequently, the focus of current uses on understanding production-oriented resistance to non-productivist policies is a product of our times.

Chapter 3, "How symbols of 'good farming' develop: the historical development of 'tidy farming'", addresses a major gap in the conceptualisation of the 'good farmer' by exploring how behaviours and objects of symbolic importance in agriculture gain and/or lose significance. An historical analysis is again employed – in this case focusing on the most critical period of change when farming moved from subsistence to scientifically based commercial agriculture. By exploring the connection between historical developments in the agricultural sector and the emergence of the symbolic significance of straight lines, tidy farms, and roadside farming the chapter pieces together a new understanding of how symbolic meaning evolves. In addition, by looking at historical literature on the features of a 'good ploughman', the chapter identifies how innate (natural), embodied (through practice), and knowledge-based (through learning) characteristics combine to exhibit the symbols of a good ploughman.

Returning to contemporary research, Chapter 4, "Theorising the 'good farmer': from a common sense category to an analytical construct", explores the use of 'good farmer' approaches in contemporary research. Here we review the diverse range of theoretical framings. While these theorisations share a similar objective – to address the cultural dimension of farming – they have different implications in terms of how culture and social interactions are articulated and how social change is integrated. We contend that highlighting the strengths and weaknesses of the framings is a much needed development and should result in more reflexive uses of the 'good farmer' approach. From the initial inspiration from symbolic interactionism, the theorisation of the 'good farmer' has mostly settled within the Bourdieusian theories of capitals and habitus. However, other recent works offer alternative useful perspectives. The chapter reveals that two key dimensions are at stake in this discussion: cultural transmission and socialisation and the relation between farmers' identity and agricultural practices.

Continuing the theoretical discussion, Chapter 5, "Morality and the 'good farmer'", addresses a largely ignored dimension of the literature's handling of the 'good' in the good farmer. Goodness implies normative assumptions and, thus, moral implications. Indeed, most often, the 'good farmer' literature, following a phenomenological inquiry, adopts a rather agnostic stance towards what is actually good or bad within competing definitions of the 'good farmer'. We argue, first, that the moral dimension is unavoidable in discussions over who is a 'good farmer', particularly in a time of environmental crisis; and second, that this integration strengthens the 'good farmer' approach in three ways. First, we can more clearly distinguish between the connections within competencies (good at) and goodness (good). Second, as researchers we learn to be more reflexive about the tensions between the moral and the sociological. Third, we can more clearly map the relationship between cultural understandings of goodness and ecosystem measurements of sustainability, productivity, and health.

Chapter 6, "The gendered 'good farmer'", draws on the literature on women in agriculture and on LGBTQ farmers to underline the strong gendered presumptions in typical definitions of the 'good farmer' (e.g. implicitly understood in terms of heterosexual masculinity). Family farming is tied to a strongly gendered division of roles, which is well documented. Without opportunities to engage in professionally articulated masculine symbols, women have negotiated their own symbols of what it meant to be a good farm wife. However, the notion of the 'good farm woman', as parallel to the 'good farmer', has not yet been clearly articulated. In this chapter, we reassemble elements from the literature to identify those symbols of the good farm woman and describe her evolution through the modernisation process of agriculture that arguably led to its masculinisation. The integration of gender allows us to open a discussion on alterity in family farming and to ask if it is actually possible to be a 'different' good farmer.

In Chapter 7, "The 'good farmer' in communities of practice", we return to the issue of symbolism, in this case exploring how farming identities are formed and shaped within dispersed communities (i.e. peer groups not constrained to neighbourhood relations). Here we are not looking at the development of symbolic meaning per se, but how these developments differ within different farming communities – the diversity of 'good farmer' meanings. We consider three specific communities of practice: organic farmers, new entrants to farming, and pedigree livestock producers. We then consider recent research into how non-farmers understand and utilise standards of good farming, particularly veterinarians and members of the public. In doing so, we delve deeper into the role of learning, and the associated implications for an understanding of 'bad farming' within a contemporary rather than historical context.

In the final chapter, Chapter 8, "The 'good farmer' of the future", we look of the future of the 'good farmer' – both in terms of the future of the 'good farmer' in agriculture him/herself given the changing nature of agriculture, and, alongside this, future developments of the concept of the 'good farmer' and its application. Will the 'good farmer' continue to be the centre of resistance to future agricultural developments? Can we use the concept of the 'good farmer' to help build sustainable food systems? As technological developments render human skill and knowledge increasingly irrelevant, will there even be such a thing as a 'good farmer' in the future? At the same time, we speculate on how we can advance the conceptualisation of the 'good farmer' in the future, what research methods we should apply, and how we can extend the cultural reach of the 'good farmer' to account for the greater diversity in agriculture today. In this way we sum up our perspective on the issues that arose throughout the book, lay out the key questions that emerged or are likely to emerge in the future, and, we hope, provide a trail for future researchers to follow in their applications of the 'good farmer' concept.

References

Allison, L. (1996). On planning a forest: Theoretical issues and practical problems. *Town Planning Review, 67*(2), 131–143.

Barnes, A. P., Moxey, A. P., Vosough Ahmadi, B., and Borthwick, F. A. (2015). The effect of animal health compensation on 'positive' behaviours towards exotic disease reporting and implementing biosecurity: A review, a synthesis and a research agenda. *Preventive Veterinary Medicine, 122*(1–2), 42–52.

Berlan Darqué, M. (1988). The division of labour and decision-making in farming couples: Power and negotiation. *Sociologia Ruralis, 28*(4), 271–292.

Billaud, J.-P. (1991). Les agriculteurs et leurs systèmes de représentation. In *L'agriculture familiale. Une réalité polymorphe* (Vol. 1, pp. 123–155). Paris: L'Harmattan.

Brown, I., Castellazzi, M., and Feliciano, D. (2014). Comparing path dependence and spatial targeting of land use in implementing climate change responses. *Land, 3*(3), 850–873.

Burton, R. J. F. (1998). *The role of farmer self-identity in agricultural decision making in the Marston Vale Community Forest.* Unpublished PhD Thesis. DeMontfort University, Leicester.

Burton, R. J. F. (2004). Seeing through the 'good farmer's' eyes: Toward developing an understanding of the social symbolic value of 'productivist' behaviour. *Sociologia Ruralis, 44*(2), 195–215.

Burton, R. J. F., and Paragahawewa, U. H. (2011). Creating culturally sustainable agri-environmental schemes. *Journal of Rural Studies, 27*(1), 95–104.

Campbell, H. (2000). The glass phallus: Pub(lic) masculinity and drinking in rural New Zealand. *Rural Sociology, 65*(4), 562–581.

Caputo, J., and Butler, B. (2017). Ecosystem service supply and capacity on U.S. family forestlands. *Forests, 8*(10), 395.

Cardoso, C. S., von Keyserlingk, M. A. G., and Hötzel, M. J. (2016). Trading off animal welfare and production goals: Brazilian dairy farmers' perspectives on calf dehorning. *Livestock Science, 187*, 102–108.

Chan, K. M. A., Anderson, E., Chapman, M., Jespersen, K., and Olmsted, P. (2017). Payments for ecosystem services: Rife with problems and potential – for transformation towards sustainability. *Ecological Economics, 140*, 110–122.

Cloke, P. (1997). Country backwater to virtual village? Rural studies and 'the cultural turn'. *Journal of Rural Studies, 13*(4), 367–375.

Collins, A. L., Zhang, Y. S., Winter, M., Inman, A., Jones, J. I., Johnes, P. J., Cleasby, W., Vrain, E., Lovett, A., and Noble, L. (2016). Tackling agricultural diffuse pollution: What might uptake of farmer-preferred measures deliver for emissions to water and air? *Science of the Total Environment, 547*, 269–281.

Droz, Y., and Miéville-Ott, V. (2001). *On achève bien les paysans. Reconstruire une identité paysanne dans un monde incertain.* Chêne-Bourg and Genève: Georg.

Emery, S. B. (2010). *In better fettle: Improvement, work and rhetoric in the transition to environmental farming in the North York Moors.* Unpublished PhD. Durham University, UK, Durham. http://etheses.dur.ac.uk/379/

Fairweather, J. R., and Keating, N. C. (1994). Goals and management styles of New Zealand farmers. *Agricultural Systems, 44*(2), 181–200.

Fishbein, M., and Ajzen, I. (1975). *Belief attitude, intention and behaviour: An introduction to theory and research.* Reading, MA: Addison-Wesley.

Forney, J. (2016). Blind spots in agri-environmental governance: Some reflections and suggestions from Switzerland. *Review of Agricultural, Food and Environmental Studies, 97*, 1–13.

Gasson, R. (1973). Goals and values of farmers. *Journal of Agricultural Economics, 24*(3), 521–542.

Gasson, R., and Winter, M. (1992). Gender relations and farm household pluriactivity. *Journal of Rural Studies, 8*(4), 387–397.

Haggerty, J., Campbell, H., and Morris, C. (2009). Keeping the stress off the sheep? Agricultural intensification, neoliberalism, and 'good' farming in New Zealand. *Geoforum, 40*(5), 767–777.

Hansson, A., Pedersen, E., and Weisner, S. E. B. (2012). Landowners' incentives for constructing wetlands in an agricultural area in south Sweden. *Journal of Environmental Management, 113*, 271–278.

Harrison, C. M., Burgess, J., and Clark, J. (1998). Discounted knowledges: Farmers'' and residents'' understandings of nature conservation goals and policies. *Journal of Environmental Management, 54*(4), 305–320.

Hastings, M. (1988). Farming wives as business managers. *Farm Management, 6*(8), 309–315.

Ilbery, B. W. (1983). Goals and values of hop farmers. *Transactions of the Institute of British Geographers, 8*(3), 329–341.

Isgren, E. (2016). No quick fixes: Four interacting constraints to advancing agroecology in Uganda. *International Journal of Agricultural Sustainability, 14*(4), 428–447.

Jakobsen, M. (2017). Consequences of intensive use of non-financial performance measures in Danish family farm holdings. *Qualitative Research in Accounting and Management, 14*(2), 137–156.

Kristensen, E., and Jakobsen, E. B. (2011a). Challenging the myth of the irrational dairy farmer; understanding decision-making related to herd health. *New Zealand Veterinary Journal, 59*(1), 1–7.

Kristensen, E., and Jakobsen, E. B. (2011b). Danish dairy farmers' perception of biosecurity. *Preventive Veterinary Medicine, 99*(2–4), 122–129.

Lankester, A. J. (2013). Conceptual and operational understanding of learning for sustainability: A case study of the beef industry in north-eastern Australia. *Journal of Environmental Management, 119*, 182–193.

Li, T. M. (2007). Practices of assemblage and community forest management. *Economy and Society, 36*(2), 263–293.

Lowe, P. D., Clark, J., Seymour, S., and Ward, N. (1997). *Moralizing the environment: Countryside change, farming and pollution.* London: UCL Press Limited.

Lowe, P. D., Ward, N., and Potter, C. (1999). Attitudinal and institutional indicators for sustainable agriculture. In F. Brouwer and B. Crabtree (Eds.), *Environmental indicators and agricultural policy* (pp. 263–278). Wallingford: CABI.

Mallein, P., and Cautrès, B. (1993). Les identités professionnelles des agriculteurs. In C. Courtet, M. Berlan Darqué, and Y. Demarne (Eds.), *Agricultures et sociétés; Pistes pour la recherche* (pp. 217–229). Paris: Association Descartes/INRA.

McGranahan, D. A. (2014). Ecologies of scale: Multifunctionality connects conservation and agriculture across fields, farms, and landscapes. *Land, 3*(3), 739–769.

Mead, G. (1934). *Mind, self, and society: From the standpoint of a social behaviouralist.* Phoenix edition (1967), C.W. Morris, ed. London: University of Chicago Press.

Morris, C., and Potter, C. (1995). Recruiting the new conservationists – farmers' adoption of agri-environmental schemes in the UK. *Journal of Rural Studies, 11*(1), 51–63.

Phillips, E., and Gray, I. (1995). Farming 'practice' as temporally and spatially situated intersections of biography, culture and social structure. *The Australian Geographer, 26*(2), 127–132.

Pouta, E., Grammatikopoulou, I., Hurme, T., Soini, K., and Uusitalo, M. (2014). Assessing the quality of agricultural landscape change with multiple dimensions. *Land, 3*(3).

Sandberg, M., and Jakobsson, S. (2018). Trees are all around us: Farmers' management of wood pastures in the light of a controversial policy. *Journal of Environmental Management, 212*, 228–235.

Saugeres, L. (2002). The cultural representation of the farming landscape: Masculinity, power and nature. *Journal of Rural Studies, 18*(4), 373–384.

Sawer, B. J. (1973). Predictors of the farm wife's involvement in general management and adoption decisions. *Rural Sociology, 38*(4), 412–426.

Seabrook, L., McAlpine, C., and Fensham, R. (2008). What influences farmers to keep trees? A case study from the Brigalow Belt, Queensland, Australia. *Landscape and Urban Planning, 84*(3–4), 266–281.

Stebbing, S. (1984). Women's roles and rural society. In T. Bradley and P. Lowe (Eds.), *Locality and rurality: Economy and society in rural regions* (pp. 199–208). Norwich: Geo Books.

Stryker, S. (1994). Identity theory: Its development, research base, and prospects. In N. K. Denzin (Ed.), *Studies in symbolic interactionism 16* (pp. 7–20). London: JAI Press INC.

Sutherland, L.-A., and Darnhofer, I. (2012). Of organic farmers and 'good farmers': Changing habitus in rural England. *Journal of Rural Studies, 28*(3), 232–240.

Sylvestre, J.-P. (2002). Les usages de la campagne et la dynamique identitaire des agriculteurs français. In J.-P. Sylvestre (Ed.), *Agriculteurs, ruraux et citadins. Les mutations des campagnes françaises* (pp. 55–70). Dijon: Educagri/CNDP.

Tautges, N. E., Goldberger, J. R., and Burke, I. C. (2016). A survey of weed management in organic small grains and forage systems in the Northwest United States. *Weed Science, 64*(3).

Tsouvalis, J., Seymour, S., and Watkins, C. (2000). Exploring knowledge-cultures: Precision farming, yield mapping, and the expert-farmer interface. *Environment and Planning A, 32*(5).

Van der Wal, R., Miller, D., Irvine, J., Fiorini, S., Amar, A., Yearley, S., Gill, S., and Dandy, N. (2014). The influence of information provision on people's landscape preferences: A case study on understorey vegetation of deer-browsed woodlands. *Landscape and Urban Planning, 124*, 129–139.

Whatmore, S., Marsden, T., and Lowe, P. (Eds.). (1991). *Rural Enterprise: Shifting perspectives on small-scale production*. London: David Fulton Publishers.

White, S. S., and Selfa, T. (2013). Shifting lands: Exploring Kansas farmer decision-making in an era of climate change and biofuels production. *Environmental Management, 51*(2), 379–391.

Williams, D., Lloyd, T., and Watkins, C. (1994). *Farmers not foresters: Constraints on the planting of new farm woodland*. Working Paper 27, University of Nottingham, Nottingham.

Wilson, G. A., and Burton, R. J. F. (2015). 'Neo-productivist' agriculture: Spatio-temporal versus structuralist perspectives. *Journal of Rural Studies, 38*, 52–64.

Wolpert, J. (1964). The decision process in spatial context. *Annals of the Association of American Geographers, 54*(4), 537–558.

2

THE ORIGINS OF THE 'GOOD FARMER'

The husband is he, that to labour doth fall,
The labour of him I do husbandry call:
If thrift, by that labour, be any way caught,
Then is it good husbandry, else it is naught.

(*Tusser, 1580, p. 16*)

The 'good farmer' has been many different things to many different people over the centuries. As we emphasize throughout the book, the concept of the 'good farmer' and what defines a 'good farmer' are not static but constantly vary as agricultural practices, the structure of farming, the social role of the farmer in society, and so on, vary over both time and space. In the earliest English literature, no mention is made of the 'good farmer', but of the 'good husbandman' where husbandry extended beyond agricultural practices, to the governance of a household – reflecting obligations to both the family and, almost peripherally, the farm. In fact, the term 'good husbandry' was used both with specific reference to managing a farm (e.g. Fitzherbert, 1534; Tusser, 1580), and reference to managing a family outside of the farm. De la Primaudaye (1589, p. 498), for example, defined good husbandry as simply attending diligently to the household through "industry in getting goods, and discrete government in spending them to good purpose". At this stage, the term 'farm' was not commonly used to refer to an agricultural holding. Until the 16th century, 'to farm' meant 'to rent' (from the Anglo-French term *ferme*) and, as such, the medieval farmer was not an agriculturalist, but a debt collector authorised to collect taxes, customs, and other duties.

It was in the 16th century that land began to be referred to as a 'farm' and the practice of managing that land 'farming', while the land manager him/herself remained a 'husbandman' (Merriam-Webster Dictionary, 2018). However, towards the end of the 16th century the term 'farmer' began to be used to refer to persons cultivating 'farm' land (e.g. Gerarde, 1597; Plat, 1594). Initially 'good farmer' and 'good husbandman' were used interchangeably with some retaining the use of husbandman (e.g. Markham's, 1616, *Cheap and Good Husbandry*) while others used the terms 'husbandman' and 'farmer' in equal measure – with seemingly no differentiation between the two (e.g. Surflet's (1600) translation of *Maison Rustique*). References to 'good husbandry' were common in the 18th century, occasional in the 19th century and are still used today to a limited extent. Yet from around the beginning of the 17th century, it was the 'good farmer' who emerged as the main contender.

Whether the term 'good farmer' as it appears in the English language literature is the equivalent of the 'good farmer' in other literatures is difficult to ascertain. There is some indication the term may have been generally applied. In particular, while in this chapter we focus on the English language literature (British and American), the agricultural literature across Europe (including Britain) has had strong connections since at least the 17th century, with journal articles translated and reprinted, farmers writing about the agricultures of other nations (particularly in Europe and the United States), and references to the best publications cited in extensive bibliographies. For example, Von Münchhausen (1766, pp. 9–28) in "Der Hausvater" lists German, English, French, Italian, Spanish, Dutch, and Swiss agricultural publications for German audiences, while key texts such as Duhamel Du Monceau's (1762) "Elements d'Agriculture" were translated directly, as the royal houses of Europe sought to spread knowledge of agriculture throughout their dominions. The shared destiny of Western agricultures with the rise of a productivist model in the 19th and 20th centuries is better documented (e.g. Potter, 1998) and despite its obvious specificity, the US example described in this chapter reflects broader global processes.

The importance of looking at the historical development of the concept of the 'good farmer' is that it enables us to view changes in the concept over extended periods of time and understand how it has been applied over the centuries. While the 'good farmer' today is most often associated with productivist conventional agriculture, our travels through the historical past illustrate that this was not always the case. Rather it depicts a 'good farmer' who is constantly changing as well as a term that is appropriated and re-appropriated by various interest groups as they struggled to shape agriculture and the farmers themselves in the image of their own beliefs and desires. Together with Chapter 3 – which examines the development of symbols of 'good farming' using a similar historical perspective – it enables the reader to understand who the good farmer is and where he/she emerged from.

Early notions of the good farmer: the good farmer in Britain, 1500–1800

Literature on agriculture from before the 18th century is scare, but literature on the notion of 'good farming' or the 'good farmer' is scarcer still. Printed accounts of how farming should be conducted (in English) first emerged in the United Kingdom in the 16th century, with three accounts in particular standing out[1] – those of John Fitzherbert (ca. 1460–1531), Thomas Tusser (ca. 1524–1580), and Barnabe Googe (ca. 1540–1594).

As the first of these, Fitzherbert's main work, the *Boke of Husbandry*, was published in July 1523 and reportedly "throws considerable light on the state of farmers in those days", presenting a vision of a yeoman farmer as an independent man with a plain living drawn from subsistence farming (McDonald, 1908, p. 14). Fitzherbert's work (which can be found reprinted and annotated by Skeat, 1882) is a manual for farmers, and covers three main areas considered important to good farming: namely, practical aspects of farm management such as the construction and operation of farm tools, veterinary skills, and management advice for conducting fieldwork; advice for the management of the farm household targeted specifically at the farmer's wife; and the religious duties of the farmer along with lessons on morality interspersed with the practical advice. Fitzherbert also refers to the need to be part of the farming community. He argues that established farmers should provide young farmers the "seed of discretion", i.e. to freely provide advice on farm management to less experienced farmers "for the more that it is taken of or lent, the more it is" (McDonald, 1908, p. 20). While the entire theme of the book is good husbandry, there is little mention of the 'good farmer'. In fact, Fitzherbert's work appears to be targeted more to country gentlemen than peasants as is indicated by the use of Latin in the text, something which Skeat (1882) observes marks him out as not being a common farmer.

Thomas Tusser's book, *Five hundred pointes of good husbandry* (initially *One hundred points of good husbandry*, printed in 1557) is the most widely cited of the three early works. Various editions of Tusser's work emerged between 1557 and 1580 and, even after his death, at least 16 reprints have been made with three of these published in the 19th century (see White, 1848; Mavor, 1812; Payne and Herrtage, 1878). At the time of his death Tusser owned two small farms and, while evidence is scarce, Mavor (1812) suggests that, although he later lived in the cities of Cambridge and London (where he died), it was probable that during his life he was directly involved in practical farming. Unlike Fitzherbert's effort to present his advice as clearly as possible, Tusser's work is in verse – reportedly to ensure that the doctrines are taken up more effectively (Stillingfleet, undated, cited in Mavor, 1812). This desire to promote his doctrine may also be the reason Tusser chose to emphasize the notion of 'good farming' versus 'bad farming' throughout the publication. As with Fitzherbert, Tusser offers lessons in agriculture but, through employing the 'good farmer' rhetoric, offers a stark choice between profitability and spiritual reward as a 'good farmer', or failure and damnation as a 'bad farmer'. His advice has a number of components.

First, in the same spirit as Fitzherbert's manual, Tusser outlines the practical aspects of being a farmer in terms of farm management, laying out the tasks in calendar form. For example, "October's abstract" states: "First barley or rye, then peas by and by. Then fallow for wheat, is husbandry great." Here 'good farming' refers to superior management practices, simply emphasizing what is best to do in the profession of agriculture. Do this, he says, and you will be a 'good farmer'.

Second, Tusser reveals a strong belief that 'good farming' is being part of a community and working with others in agriculture. In a number of places, he denounces the slandering of others, while, at the same time, promoting the maintenance of good relations with neighbours through hospitality. Tusser further suggests that lending to neighbours in their time of need "wins love of thy neighbour, and credit doth breed" (p. 22), which suggests an understanding that being a good neighbour has rewards in terms of the development of social capital.

Third, Tusser, who was deeply religious, contended that a 'good husbandman' had virtues that differed from a 'bad husbandman'. In particular, a section titled "The description of an envious and naughty neighbour" outlines a cornucopia of "naughty neighbour" sins including greed, vengefulness, pride, boastfulness, being quick tempered, deviousness, being disingenuous, and engaging in gossip. In addition, Tusser frequently refers to the good husbandman's obligations to God, for example, that he should "seek to God for remedy, for witches prove unluckily" and "pray God a good harvest to send".

Finally, and similarly to Fitzherbert, Tusser emphasizes the importance of the farmer's wife in good husbandry. Although her many tasks (which Tusser outlines at length) were associated with the management of the household rather than the fieldwork, for Tusser they appear as critical to good farming as the role of the male farmer. He observes:

> Of husband, doth husbandry challenge that name,
> Of husbandry, husband doth likewise the same:
> Where huswife and huswifery joineth with these,
> There, wealth in abundance is gotten with ease.
>
> (Tusser, 1580, p. 16)

Tusser also lays out "A comparison between good huswifery and evil" (p. 184) in which, rather than a list of tasks for the 'good housewife', he lists virtues (or ill-virtues) including slothfulness, gossiping, spendthrift, and pride as indicative of someone who is not a good housewife.

Constructed from the works of authors ranging from the classical era onwards and organised as an imaginary dialogue between four persons, Barnaby Googe's (1578) *Whole Art of Husbandry* provides another perspective on good farming – one that appears to be largely literature based and is in even more obscure verse than that of Tusser. This publication consists of Googe's translation of the work of the German Conrad Heresbach, himself interpreting classical scholars of agriculture – although Googe notes he "altered and increased his work with mine own readings and observations" (Googe, 1578, p. iii). Given the origins of the publication and the lack of clear attribution throughout, it is often difficult to discern which period in history and from which author the original ideas emerged. While the text is somewhat more convoluted than that of Fitzherbert or Tusser it also provides advice to farmers on how to manage their farms and emphasizes the importance the farmer's wife, noting:

> I *confer with my wife* or servants of husbandry, appointing what I will have done: if my bailiff have any thing to say: if anything to be bought or sold: for a good husband, as Cato says, must rather be a seller than a buyer.
>
> (p. 4)

Religion is also an important factor in Googe's account. He notes before going to bed,

> Sometimes, (especially in winter) after supper, I may tell something out of the holy scripture, or else some pleasant prose, so that it may be honest and godly, and such as may edify.
>
> (p. 4)

These three accounts of farming in the 16th century present a picture of the good farmer that, while similar to today's 'good farmer', show some distinct differences. First, while the farm wife clearly had gender defined roles, the accounts suggest that good farming (or, more correctly, good husbandry) historically was an integration of predominantly male production roles and the female role of farm household management – such that being a good farmer necessitated the involvement of a 'good housewife'. F.B. (1672, p. 19) in *The Office of the Good Housewife* observes that "The condition and state of the housewife or dairy woman is of no less care and diligence than the office of her husband" but stresses that the farm wife does not perform field work but is "tied to the matters of the house" and, similar to Fitzherbert and Tusser, provides a long list of the roles for the 'good housewife'. Some of these roles are more commonly male roles today, such as the responsibility for the management of breeding livestock – i.e. the housewife would "take care that the cows be provided with a bull" (F.B., 1672, p. 25). The greater involvement of the wife historically makes sense from the perspective that many of the manufacturing tasks (e.g. making linen or cheese) were performed on the farm and thus a successful farm required a higher degree of labour and management input. We more directly address contemporary issues related to gender and the good farmer in Chapter 6.

The second major difference is the role of morality and godliness in the concept of the good farmer. In addition to agricultural tasks, good farmers were supposed to be virtuous in the eyes of God. In part this may be attributable to references to the nobility of the farming profession in both the Bible and the classical literature authors. For example, Googe (1578) observes that historically the "most mighty kings and emperors" bore a "singular affection for husbandry" while emperors were known to come "from poor herdsmen to the imperial vignette" (p. 5). Googe reports that, at that time, "as oft as they would give a man the name of an honest man, they would call him a good husband, comprehending in him in that name as much commendation, as they could give him" (p. 5). Others held similar views based on religious doctrine. Markham (1635) in *The English Husbandman*, for example, extols the high virtues of the husbandman in observing:

> Let every man understand that this title of husbandman is not tied only to the ordinary tillers of the earth, such as we call husbandmen; in France peasants, in Spain basonyans, and generally the clout-hoo. No, *they are creatures of a better creation*, for as Adam was the first husbandman so all the sons of Adam (even from the crown to the cottage) *cannot assume a greater a better or richer title than to be a good husband.*
>
> *(p. 4)*

In general, early characterisations of the good farmer differed dramatically from those of the conventional productivist farmer identities that dominate today (see Burton and Wilson, 2006). While innovation, efficiency, and increasing production were primary goals, at this time the 'good farmer' was a more integrated character – a producer of food, yes, but also part of a family, a community, and a religion. In a sense, good husbandry of the land was integrated with 'good husbandry' in the other sense it was used, that of maintaining the household economy and the caring obligations to the family. Being good at agriculture was essential, but one could not be a 'good farmer' without also being a good husband.

Why 'good farmer' rather than 'good farming'?

Frequent use of the term 'good husband' and 'good farmer' throughout the 16th to 19th centuries raises the question of what was the intended purpose of creating a 'good' and/or 'bad' farmer? Discerning between good and bad agricultural practices is obviously important to emphasize the difference between patterns of behaviour likely to result in positive outcomes for land managers (i.e. outcomes espoused by the author), and those that do not. However, if this were the sole purpose, it would be possible to present it as 'good farming', rather than creating the 'good farmer' and his/her counterpart the 'bad farmer.' The fact that reference is frequently made to 'good husbandry' or 'good farming' illustrates that it is easily possible to advise readers how to improve their farming practices without personifying the concepts, so why is it necessary to create the good or bad farmer?

From one perspective the 'good farmer' appears to be a rhetorical tool for imprinting the message of the book. Severe criticisms act as a warning and emphasize the idea that you do not want to be identified as that kind of farmer. Plat (1594, p. 9) for example, draws a distinction between the "learned husbandman master Bernard Palissy" who "wrung out of the bowels of the earth" his knowledge of agriculture and whose publications serve as the basis of Plat's work, and the "ignorant farmers (who) may also glean with them a few loose and scattered ears to make so much bread of, as may relieve their hungry bellies." Where the concept of the 'good farmer' is used, it is often to emphasize that the practice being advocated is a critical one. For example, Anon (1829) advocates for increasing manure usage by observing

> Manure is the great sinew of agriculture, as money is of war; and the making the best of every advantage or opportunity for increasing the quantity of it, is one of the most prominent traits in the character of a good farmer.
>
> (p. 39)

Thus, by making certain behaviours indicative of the 'good farmer' authors emphasize certain traits over others while presenting the individual with a stark choice – be good or bad. In other cases, however, it may simply be used for poetic purposes as is particularly evident in Tusser's work, which is written in verse. For example, in the case "Good husband and huswife, will sometime alone, make shift with a morsel and pick of a bone" (Tusser, 1580, p. 175), the terms 'husbandry' and 'huswifery', while equally applicable, would break the meter of the verse. Another possible reason is that the concept of the 'good farmer' reflected the structure of information transfer in farming communities prior to the development of experimental agriculture and formal institutions – which did not exist to any great degree until the first half of the 19th century (Loudon, 1839). In a profession with widespread illiteracy and limited access to validated scientific evidence, success was often measured by individual reputations, with 'good farmers' serving as a focus for learning and information transfer in their communities. Early writers sought to justify their own agricultural knowledge by identifying 'good farmers' as the source. For example, in reference to a response to the Bath Agricultural Society, it is noted that the answers "were given him by a very good farmer, and approved by all who had seen them" (Anon, 1783, p. 57). Consulting with good farmers was also an important form of validation for those with enquiries on farm management. Lisle (1757), for example, observes:

> At Christmas-time (anno 1700) several good farmers being with me, I was enquiring for peas and barley for seed. They replied, that the housing of corn had been so good this year, the buying of seed might be ventured on the earlier, else they used not generally to buy their seed-barley, nor seed-oats, but just before sowing-time, lest they should smell by heating, and so not grow.
>
> (p. 154)

More recent literature on the 'good farmer' suggests being identified as a good farmer plays an important social role in farming communities, in that it enables farmers greater access to social capital – common resources obtainable from other members of the community (Sutherland and Burton, 2011). In the historical case being identified as a 'good farmer' may have had strong economic implications for the farmer him/herself. For example, Robert Bakewell from Dishley Grange (Leicestershire, UK) built up a livestock farm in the mid-1700s that was identified as a "model for all practical farmers to study" (Anon,

1842, p. 85). However, for Bakewell, who generated most of his income from livestock breeding, this reputation of being a 'good farmer' undoubtedly assisted his highly reputable and successful stud business.

Finally, in recognising that 'good farmers' formed the centre of information transfer within their communities, early authors may have employed the concept to encourage wider change in agriculture by seeding 'good husbandmen' among the community. Plat (1594) for example, observes that farmers would marle (a form of fertiliser) their lands and

> in so doing thou shalt show thyself a good husbandman and become a pattern to all they slothful neighbours, whereby they shall be forced to imitate thy good example.
>
> *(p. 26)*

Certainly, at a time when formal agricultural education was scarce and demonstration agriculture in the United Kingdom relatively ineffective (Burton, 2020), encouraging a generation of 'good farmers' to aid the development of farming communities was probably the most effective a means available for promoting development.

The changing identity of the good farmer

Historical references to the 'good farmer' generally recognised only a single farmer 'type', whereas now, when we think of the good farmer, we conceive that there may be different cultural categories of farmer (e.g. Sutherland and Darnhofer's, 2012, distinction between conventional and organic farmers). One of the first observations of different farming types can be found in William Marshall's (1778) classification of farmers according to their dispositions and practices.[2] Typologies distinguishing between farming types or cultures have become relatively common in the last few decades, for example, in the form of Ploeg's widely applied "farming styles" approach (Ploeg, 1994) or through numerous bottom-up data driven typologies (e.g. Bakker and van Doorn, 2009; Davies and Hodge, 2007). Marshall's main concern in deriving this typology was to identify which types of farmers may be most useful for improving the condition of English agriculture. In the typology farmers were divided into three principal classes: the aboriginal farmer, the scientific farmer, and the aerial farmer.

Aboriginal farmers, Marshall (1778) contended, can be divided into five main types:

The illiberal sloven – The illiberal sloven was "A sullen, designing, incommunicative being, who holds his profession as a family-secret, though his management be execrable, his fields are foul, his crops wretched, his live-stock pitiable; his whole life is a scene of cunning, toil, poverty and wretchedness." *(p. 8)*

The illiterate economist – An illiterate economist was similar to the illiberal sloven, being "Equally disingenuous, equally incommunicative" but he "has more cunning, and is a better farmer. He keeps his land tolerably clean, and in tolerable heart; his crops, and every thing around him are tolerable and he gets rich in proportion to his rent, taxes, and luck" (p. 8).

The ape-gentleman – the ape gentleman is a farmer who has made sufficient capital in agriculture to consider himself a gentleman, and adopts all the trappings of a gentleman farmer (wine consumption, horse riding, lavish entertaining, etc.) – but the family "is extravagant without gentility" (p. 8). His good fortune creates indolence and pride and, in the end, management of the farm becomes lax and the business declines.

The substantial – The substantial farmer is a core farmer in Marshall's aboriginal classification. He reportedly "springs from a long line of economists and industrious housewives. He is possessed of a fortune independent of farming; he is the Lord of the Village he lives in; and deals out comfort or oppression to his poor neighbours." In terms of their social position, "his whole family is a pattern of rural decency, and their manners a relic of ancient hospitality . . . though he is a customist, and, of course, cannot communicate his general management, he knows the customs and usages of the Country he lives in, and can obviate a thousand petty difficulties" (p. 9).

The gentlemanly – the gentlemanly farmer is the educated son of a substantial farmer. While he is seen as a liberal, Marshall contends he will never become a scientist, noting "He may improve but can never perfect; because, he is prejudiced by the custom of his ancestors, and the Country he happened to be bred in, whose errors he has not only received as immutable truths, but his whole life has been employed in the practice of them" (p. 10).

The *Scientific farmer* appears to be the ultimate 'good farmer' of Marshall's typology – one which Marshall believes all others should emulate. Consisting of a single type, the scientific farmer is one who "believes that the customists collectively are far advanced on the road of perfection; but he knows nothing in agriculture, which proceeds not directly from analysis, experiment, and observation" (p11). The scientist is reportedly concerned with desires to optimise production – to see each soil type "occupied by the Vegetable which affects it and, collectively, by such as are most immediately subservient to the natural necessities of man" (p. 12).

Aerialist farmers, on the other hand, are those that are always moving from one thing to the other, chasing the "chimeras of vast improvement and vast profit" (p. 12) based on knowledge from books rather than experiment or experience. However, aerialist farmers do not have the understanding of the scientific farmers, nor the wisdom of the substantial or gentlemanly farmers. "He is constitutionally volatile, speculative, and credulous, and habitually bookish. He has acquired a smattering of the lighter branches, the twigs of human knowledge and apes genius, though void of penetration and judgment" (p. 12).

Marshall's typology portrays some interesting changes in the notions of good farming that occurred in the second half of the 18th century, presenting a more nuanced view of the 'good' and 'bad' farmer – ranging from the aerialist farmers who ought to be reformed or 'annihilated', to the scientific farmer who all others ought to emulate. Farmers for Marshall were no longer simply 'good' or 'bad', but came in different shades of 'good' and 'bad', identifiable by their different but equally flawed approaches to farming. The scientific farmer, largely devoid of customs and focused on optimising production on the basis of scientific principles, appears to be a precursor to the modern productivist farmer. This reflects the transition occurring at the time from tradition-based agriculture to experimentation-based agriculture – from small-scale self-sufficient production to production for markets to feed the growing cities in Europe. A noticeable turn in the literature can be found from the mid-18th century onwards from a period where advice was either customary or anecdotal (providing references to good farming practices) to one where scientific principles and experimental agriculture formed the basis of evidence for improved practices. As the 18th century came to a close, the good farmer was decreasingly portrayed as the community oriented godly farmer supported by his industrious wife, but as a farmer scientist whose main concern was to improve the productivity of agriculture to fill the world with "universal plenty" (Marshall, 1778, p. 11).

The good farmer in the United States – new land, new peoples, new concepts.

Through the early part of the 19th century scientific principles played an ever-increasing role in agriculture while traditions and customs were declining. Agriculture was changing from the use of simple implements to being reliant on increasingly complex machinery and, as we outline in Chapter 3, this meant considerable changes to the role of farmers and the notion of good farming. Nearing the mid-point of the century the changes had become so dramatic that Loudon (1839, p. 1206) in the "Encyclopaedia of Agriculture" contended that publications prior to 1830 were "of very little value" from a scientific perspective but were "to be considered as historical documents of the progress of opinions and practices". Britain was no longer the only Anglo country developing its agriculture at a rapid pace, with the colonies in Australia and New Zealand, desperate to both feed their growing settler populations and develop agricultural industries for export. This was a new type of agriculture, largely devoid of tradition and the land ownership systems of the past, where individual farmers made their own rules and carved out their farms from land taken from indigenous populations.

The 19th century also saw the United States begin to play an important role in defining the 'good farmer', in particular, between the US Civil War (concluded in 1865) and the US entry to World War II (1941), a point at which we argue that the productivist mindset was set as the model of global agricultural policy. As with the various iterations of 'farmer' and 'good' in the UK examples, we see the use of 'successful', 'better', and 'master' in some of the following examples, but the ideal of what is 'good' remains central. In the United States, the connection between farmers and citizenship started with Thomas Jefferson – farmer, drafter of the Declaration of Independence, and second president. In a letter to John Jay, Jefferson (September 23rd, 1785) wrote

> Cultivators of the earth are the most valuable citizens. They are the most vigorous, the most independent [*sic*], the most virtuous, and they are tied to their country and wedded to its liberty and interests by the most lasting bands.

As historian Don Worster (1993, p. 87) argues, "We have been taught from Thomas Jefferson's day on down that what is good for the farmer is good for America". While Jefferson retreated from an early assertion that the country would be better off with ever more farmers, this Jeffersonian agrarianism still captures American farmers' imaginations. In regard to what a good farmer is specifically, Danbom (1991) articulates two major strains of agrarianism (rational and romantic) that both believe they are contributing benefits to the development of the country and themselves as citizens. While agrarianism gets at part of what it means to be a good farmer in the United States, the rest of this chapter will highlight three examples of good farmers from the period between circa1870 and World War II – those of freed slaves, the development of demonstration and extension systems, and the increasing industrialization and professionalization of farming in the 20th century. These stories, in parallel with the UK examples, help us comprehend the interrelated development of industrialisation in farming and the evolution of a productivist 'good farmer' ideal.

Making good citizens: African Americans as good farmers

The emancipation of former slaves created multiple challenges for US society, many of which remain unresolved (White, 2018). Many African Americans moved North rather than returning to Africa, while others remained scattered across the South and became sharecroppers, an arrangement whereby land that once was intended to be freely distributed to former slaves (the origins of the 40 acres and a mule mythology) was now being farmed in a tenancy agreement. While England had seen some move towards tenants achieving independence and self-determination (Haggard, 1899), the same was not true for African Americans in the United States. This struggle and outright racism developed into the sharecropping system that came to dominate the American South (Daniel, 2013; Seavoy, 1998; White, 2018). As W. E. B. Du Bois (1903) wrote:

> among such conditions of life there are few incentives to make the labourer become a better farmer. If he is ambitious, he moves to town or tries other labour; as a tenant-farmer his outlook is almost hopeless, and following it as a makeshift, he takes the house that is given him without protest.
>
> *(p. 211)*

At the same time African American cooperatives sprouted across the South, utilising farming as a form of resistance to economic and political domination including the emergence of white supremacist groups like the Ku Klux Klan (White, 2018, p. 14). The recent freedom of so many people with specific skills in agriculture led to a conundrum in the agricultural press. On one hand the (overwhelmingly white) press made examples of a few talented African American farmers to demonstrate the benevolence of white mentorship using 'this could be you' journalism to inspire both other African Americans and 'lazy' white farmers. On the other hand, there emerged a black agrarianism – notably through Frederick Douglas and George Washington Carver along with numerous slave narratives (Glave, 2010; Smith, 2004) – that aimed for a Jeffersonian ideal where farming offered a path to full citizenship for the newly freed slaves. Smith (2004, p. 270) argued that black agrarianism made a pitch that "secure property rights create economic and emotional incentives for good stewardship, while an equitable labour system creates a respect for labour that encourages good farming". In the agrarian tradition, then, having secure access to land "in turn leads to careful stewardship" or good farming (Smith, 2004, p. 283). Turning disadvantaged or oppressed groups into good citizens by making them 'good' farmers was not unique to African Americans in the United States, but was also a strategy applied to colonised Ireland as well as indigenous American tribes in the mid-19th century (Burton, 2020).

As one might imagine, much of the commentary by white agricultural writers discounted the skill of African Americans as farmers (e.g. 'Successful Negro Farmer', 1886) despite a long history of contributing admirably and creatively to the strength and emergence of Southern agriculture (Carney, 2001). Sharecropping, particularly around cotton, saw former slaves offered parcels of land on exhausted soil, often from former plantation owners now not in residence who then took much, if not all, of the harvest each year. The exceptions to this system – those lucky to be working a fertile piece of land and lacking a vengeful landowner/landlord – were celebrated in genre pieces that highlighted that the farmer should be seen as a 'credit to their race'. For example, the *Goldsboro Messenger* noted in 1886 about Scipio Williams:

> Williams, a coloured man, farming on the plantation of Capt. W. S. Ellerbe . . . is a phenomenal negro in many respects, a credit to his race and a living refutation of the idea that there is no money in farming.
>
> *(p. 7)*

The article goes on to speculate on what it was that made Scipio a good farmer including succeeding "by dint of hard blows and indomitable energy", being thrifty, being independent, feeding his family, and that he "never mixes in politics or meddles with other people's business" – a thinly disguised reference to knowing his place in a white society. As with the case of Irish peasant farmers under colonial rule, the notion of a 'good farmer' was associated with a sedate population controlled by their dedication to the land and obligations to their families in such a way that they were unlikely to cause problems for the dominant classes. Yet despite the white narrative of a good black farmer as never mixing in politics, for many former slaves, their narratives expressed hope to enrol "democratic agrarianism to establish blacks' rights to the land and citizenship" (Smith, 2004, p. 277).

The 'good farmer' as a transformative social force: George Washington Carver and Seaman Knapp

The late 19th and early 20th centuries were a period of profound transition and contradiction, with agricultural progress being defined as both backward looking and forward looking. 'Backward' progress was seen in the embrace of soil buildup and ecological awareness about waste embodied by Carver, Albert Howard, Gabrielle Matthaei, Louise Matthaei, Rudolf Steiner, Lady Eve Balfour, and J. I. Rodale. 'Forward' progress would come to be defined as productivism and involved the development of an agriculture dependent upon chemicals, artificial fertiliser, and ever-larger machinery as was occurring in the United Kingdom and across Europe. To help understand what a good farmer was in the late 19th and early 20th century United States, we turn briefly to George Washington Carver's time at the Tuskegee Institute. Tuskegee was set up in the wake of the Civil War to extend practical skills to African Americans in the Macon, Alabama, area. Once George Washington Carver (an African American) joined Tuskegee in 1896 he set about researching, writing, and speaking about all of the ways in which African American farmers could improve the soil and, thereby, their lives. His bulletins were filled with the latest technological innovations, recipes, and crop suggestions indicative of Carver's idea of what a good farmer could be, given such great limitations on land, credit, and technology (Hersey, 2011; White, 2018). Carver's agrarianism tapped into the ideals of Jeffersonian connections between good farming and good citizens. While abandoning some of the boosterism for new technological advancements (simply because his constituents could never afford them – Hersey, 2011), Carver advocated for the restoration of the highly depleted southern soil. Carver went so far as to write that taking care of the soil is the moral equivalent of taking care of other people. As Hersey (2011, p. 142) describes Carver's stance: "High yields and rich profits alone did not a good farmer make."

In many ways we see the emphasis on soil over profit evident in the relationship between Carver and Knapp. Seaman Knapp met Carver at Tuskegee in 1906. Three years prior, Knapp – realising that farmers distrusted government agricultural agents because they had no economic stake in the outcomes – pioneered the US demonstration farms by convincing a private farmer in Texas to set aside 70 acres (Hersey, 2011, p. 155). Three years later Knapp visited Tuskegee to find the mobile demonstration school an inspiration. This mobile school consisted of a wagon outfitted with tools, seeds, and other materials relevant to improving not only farming but life in general through providing information on homemaking, hygiene practices, and so on. For rural Alabaman African Americans the mobile school out of Tuskegee provided Carver and others a chance to share what they had learned and promote their vision of the 'good farmer'.

While it is difficult to judge the "success" of the mobile school, two things are notable. First, what was included on the wagon certainly demonstrates what Carver and others at Tuskegee's agricultural school saw as representative of good farming: "different kinds of plows and planters, a cultivator, a cotton chopper, a variety of seeds, samples of fertilizers, a revolving churn, a butter mold, a cream separator, a milk tester, and other appliances useful in making practical demonstrations" (Mayberry, 1991, p. 87). Second, the mobile school made an impression on Knapp during his visit such that Knapp would include versions of the demonstration wagon/van in the creation of the county extension agency that Knapp would spearhead soon after. Aligning with agricultural schools and local governments, the extension system provided funding to make sure that every county had an extension office to support the farming needs. While this model become more productivist as the years went on (see Hightower, 1978), at the beginning Knapp envisioned extension agents would carry out his vision of agriculture as embodied in his Ten Commandments of Agriculture.

Knapp's Ten Commandments of Agriculture (Knapp, 1909, p. 153):

1. Prepare a deep and thoroughly pulverised seed bed, well drained; break in the fall to a depth of 8,10 and 12 inches, according to the soil, with implements that will not bring too much of the sub-soil to the surface. The foregoing depths should be reached gradually.
2. Use seed of the best variety, intelligently selected and carefully stored.
3. In cultivated crops give the rows and the plants in the rows a space suited to the plant, the soil and the climate.
4. Use intensive tillage during the growing period of the crops.
5. Secure high content of humus in the soil by the use of legumes, barnyard manure, farm refuse and commercial fertilizers.
6. Carry out a systematic crop rotation with a winter cover crop.
7. Accomplish more work in a day by using more horse power and better implements.
8. Increase the farm stock to the extent of utilizing all the waste products and idle lands of the farm.
9. Produce all the food required for the men and animals on the farm.
10. Keep an account of each farm product, in order to know from which the gain or loss arises.

What Carver and Knapp offer is a related, but divergent idea of what it meant to be a good farmer. While Carver was concerned with using farming as a way to secure a better hold on citizenship for often poor and illiterate farmers, Knapp was focused on broadly transforming rural society at a national scale, through the improvement of agriculture. For example, Knapp (cited in Martin, 1921) details how creating a new concept of good farming through demonstration would lead to major social change:

> His neighbours ask him how he produced it. He is invited to address public assemblies. He has become a man of note and a leader of the people and cannot return to his old ways. Soon there is a body of such men; a township, a county and finally a state is transformed.
>
> *(p. 17)*

This idea that society can be bettered by transforming 'bad farmers' into 'good farmers' was also evident in the origins of demonstration farming in the early 1800s when Mark Fellenburg established the Hofwyl agricultural institute in Switzerland (Burton, 2020). Whether improving the lives of former slaves or creating 'men of note' to lead counties, the notion of the creating good farmers as a transformative social force illustrates how the 'good farmer' has not always been defined solely by production or technical skills.

The industrialisation and productivisation of the 'good farmer' in the U.S.

The extension work of Knapp and Carver in the early 20th century helped to define the notion of the 'good farmer' in America through improving cultivation and management approaches. At around the same time, however, a new force that would become important for defining the role of the 'good farmer' was emerging: the introduction of lightweight tractors to replace horsepower. This was not the first time fossil fuels had been used to power agricultural implements. The development of the coal-fired steam plough (with the plough dragged along wires) in the mid-19th century enabled farmers to plough faster and in straighter lines than previously possible. The cost of establishing steam ploughing (for example, reorganising field boundaries, buying the machinery, employing skilled engineers), however, meant that steam technology was available only to wealthier farmers and landowners. The arrival of the internal combustion engine meant that lightweight mobile machinery capable of treating fields with a precision never before possible was finally available to common farmers – with the result that expectations for tidy farming increased and the role of mechanic gradually grew to become part of the farming role. Farmers finally had a source of power that, unlike the steam machinery of the past, was fast, mobile and lightweight – capable of driving onto soft fields without destroying the soil structure and doing the job of the horse without tiring.

At the same time, it was increasingly evident that agriculture was becoming a business and, with that, notions of God, community, and family being part of the good farmer identity were, in many cases, declining in favour of business acumen. Throughout the late 19th century the perceived role of farms and farmers as the bedrock of democracy also gradually diminished. With the evolution of the railroad, the telegraph, the creation of grain futures (Cronon, 1992), and the opening up of global markets, agriculture in the United States began to privilege huge landholdings with yield, not necessarily quality, seen as a lodestone for success. Consequently, what was being held up as an illustration of good farming was not small family farming conducted by farmer stewards, but industrial-scale agriculture where efficiency and profit were primary objectives. The farmer could become an industrialist, adopting the language of efficiency, concentration, and economies of scale. The logic of the industrial revolution transferred easily into agriculture – to optimise production, agriculture could be operated like a factory and provide ever greater returns on the capital invested.

In the United States this development was seen by some as both inevitable and undesirable. As Carver (1914) lamented,

> The growth of large-scale production in other industries has led very naturally to the query whether the same tendency must not show itself sooner or later in farming. That would be a pity, because it would mean, not only that the small farmer would disappear, but that the last large class of self-employed men would be eliminated.
>
> *(p. 5)*

However, strains of Jeffersonian agrarianism still held strong as a counter to the increasing reality of larger, more mechanised farms in the United States. Echoing Carver's and Knapp's emphasis on soil health, it was insisted in the Michigan Farmer (W.H.P., 1905):

> Farming whose chief aim is immediate financial returns regardless of the future is not successful farming. It is, one of the great faults of our farming. . . . The only kind of farming which can be called successful is that which attaches a bold and intelligent peasantry to the soil because they love that manner of life for its own sake more than for its financial returns, and which thus natural loves and takes care of the soil. We are trying to cultivate too much land. We need more farmers and better ones, settled on less land and better land. Therein lies the solution of many vexing questions, including falling fertility and the hired man.
>
> *(p. 469)*

About the same time, Liberty Hyde Bailey's (1915, p. 220) conclusion to his report on the Country-Life Movement in the United States provides an idealised and hopeful summation of the good farmer:

> My reader may wish to know what constitutes a good farmer. I think the requirements of a good farmer are at least four:
>
> > The ability to make a full and comfortable living from the land;
> > to rear a family carefully and well;

> to be of good service to the community;
> to leave the farm more productive than it was when he took it.

In 1915, Bailey found himself in agreement with Carver and Knapp (and some romantic agrarian strains), along with resonance with the UK emphasis on the interrelatedness of land, family, autonomy, service, and stewardship as the makings of a good farmer. In contrast to a vision of good farmers in balance with community and the global economy, World War I's conclusion led to both a period of increasing mechanisation and a related achievement of overproduction. The resultant price drop began a long farm recession that would mark the beginnings of and continue through the Great Depression. The ill-conceived notion of farming on the US plains and replacing the native prairie with monocultural cotton production combined with severe drought and windstorms created the Dust Bowl era. The resulting mortgage, farm, labour, and migration crises put a pause on more mechanised production and a focus on yield was called out in the US government. On the eve of Roosevelt's election and the New Deal that would transform economic policy in the United States, Arthur Capper (1931), a senator from Kansas, offered a scathing criticism of the predominant attitude towards farmers (and the model set up by Knapp in part):

> For generations these agencies have been urging the farmer to become a more efficient producer. He was told that the man who made two blades of grass grow where one grew before was a public benefactor – and would be rewarded for his efficiency and benefactions by getting more profits from his farming operations. . . . The agricultural colleges taught his children how to grow more to the acre on good ground; how to grow wheat on the arid and semi-arid lands that had been a few short years ago marked on the maps as the Great American Desert. The farmer responded to the advice, benefited by the instruction. He followed improved methods; he bought the latest machinery; he increased his efficiency; he became the most efficient producer of foodstuffs the world has ever seen. And what has been the result? His taxes, his living expenses, his cost of production, his transportation and marketing costs have increased as his efficiency increased, and in even larger proportions. But his reward for increased efficiency has been lower prices for his products, an increase in the mortgage on the farm, probable foreclosure of the mortgage, bankruptcy, and no place for him to go when he leaves the farm.

By this stage Knapp's original work had been turned into a major institution intent on selling a particular brand of farming called out by Capper and others. Just like the wider Agrarian Question, the systems in which farmers found themselves enmeshed encouraged one kind of farming (bigger, more capital- and technologically intensive, more corporate) at the expense of the farmers themselves. Increasingly, to be a 'good farmer' meant making farmers vulnerable to losing, firstly, their farms, and secondly, their identity as farmers. Waring and Teller (1944) outline the reality of this evolution in the midst of World War II:

> In a technological society like ours the tendencies are toward bigness, and little people are being crowded to the wall in agriculture just as in other parts of our economy. . . . The big operators and plantation owners do not want small farmers and tenants to become independent operators producing war crops. They want instead an army of surplus underpaid labour ready to work for what is offered. A continuation of this process will result in a widespread system of "factories in the fields.
>
> *(p. SM18)*

The beginning of the 20th century represents a period where ideals of agriculture and 'good farming' began to clash. Whereas in the 18th century and early 19th century the battle had been between customary agriculture and scientific agriculture – with few arguing for the virtues of customary practices – in the early 20th century the notion that small scale farming by a "bold and intelligent peasantry" focused on maintaining soil fertility was being forwarded as a competing concept of 'good farming'. The rhetoric of the 'good farmer' was no longer the sole preserve of those who sought improvements in yields and practices through capital and technology. However, the food shortages of World War II led to the emergence of state driven productivist agriculture that tipped the balance of good farming in favour of industrialisation and ever-increasing yields. Production and the distribution of surplus became paramount. Most importantly, the good farmer in the productivist period became the one who most efficiently produced material for the global marketplace. Their contribution to the country in other ways (the health of the soil, contributions to the community well-being, the family tradition, and so on) become secondary to efficiency and increasing production until the emergence of butter mountains, milk lakes, and increasing environmental degradation caused by industrial modes of production in the 1980s.

Conclusion

In this chapter we witnessed the 'good farmer' change from a family man – a man of God whose role as a husband was so intertwined with the farming role that the terms 'good farmer' and 'good husband' were used interchangeably – to one where the role has become predominantly one of production. One thing evident from the analysis is how the notion of the 'good farmer' is strongly tied to the condition of agriculture. In the 16th century subsistence farming rather than commercial farming dominated. The role of the farmer was then survival of the family as a social unit and, as a consequence, the life of the 'good farmer' revolved around wife, family, community, and God with, as we note in Chapter 3, little concern for symbols of production such as tidy farms and fields. This was not unique to that particular period of time. We witness from Knapp's efforts in the early 19th century and the story of the African American 'good farmer', that whenever the conditions for agriculture are poor and farmers struggle for survival, the role of the 'good farmer' as a community builder returns. With the growth of cities, improved transport links, and external manufacturing, agriculture became focused on supplying food and raw materials for consumption or manufacture elsewhere – and the stronger this effect, the more the 'good farmer' moved from a family to a commercial role.

The story of the 'good farmer', however, is more complex than this. The histories of the 'good farmer' presented here represent but a few of many and, as they rely on the observations of literate men (and they *were* almost exclusively men) rather than practical farmers, these histories do not necessarily represent the story from the farmer's perspective. Eighteenth and 19th century views in particular, as we observe in Chapter 1, are more often attempts to define what a good farmer is for the purpose of initiating change, rather than attempts to find out who the good farmer actually is on his/her own terms. In reality a 'good farmer' is not a literary being, but emerges organically from the interaction between farmers, society, and, in particular, the knowledge, implements, and machines that are used to work the land. As we note, one becomes a 'good farmer' by reaching a level of competence in a set of culturally significant behaviours – which could vary significantly from place to place and over time. How 'book' farmers chose to judge the good farmer was of little concern to those who had to earn their living from the land – while observing their neighbour for signs of being a 'good' or 'bad farmer' was often critical for success. But what are the signs and symbols of being a good farmer and how do they evolve? In the next chapter we address this question by exploring what it is that defines a good farmer from the farmer's perspective.

Notes

1 For a biographical account of these and other early English authors see McDonald (1908).
2 While Marshall (1745–1819) became a wealthy estate manager/owner he was the son of yeoman farmers and, according to Fussel (1949), unlike Arthur Young and other improvers, did not dismiss customary practices of common farmers as backwards and thus needing to be improved out of existence.

References

Anon (1783). Queries from the bath agriculture society. *Bath Agricultural Society, 1783*, 55–65.

Anon (1829). Manure. *New England Farmer and Horticultural Journal*, 7(5), 39.

Anon (1842). A brief sketch of Robert Bakewell Esq. *The Farmer's Magazine S2, 6*, 81–85.

Bailey, L. H. (1915). *The country-life movement in the United States*. New York: Macmillan.

Bakker, M. M., and van Doorn, A. M. (2009). Farmer-specific relationships between land use change and landscape factors: Introducing agents in empirical land use modelling. *Land Use Policy, 26*(3), 809–817.

Burton, R. J. F. (2020). The failure of early demonstration agriculture on nineteenth century model/pattern farms: Lessons for contemporary demonstration. *The Journal of Agricultural Education and Extension, 26*(2), 223–236.

Burton, R. J. F., and G. Wilson (2006). Injecting social psychology theory into conceptualisations of agricultural agency: Towards a 'post-productivist' farmer self-identity. *Journal of Rural Studies, 22*(1), 95–115.

Capper, A. (1931). What the hard-driven farmer is thinking. *New York Times,* July 26, XXI.

Carney, J. (2001). *Black rice.* Cambridge, MA: Harvard University Press.

Carver, G. W. (1914). Be kind to the soil. *Negro Farmer,* January 31, 5.

Cronon, W. (1992). *Nature's metropolis: Chicago and the Great West.* New York: W.W. Norton and Company.

Danbom, D. B. (1991). Romantic agrarianism in twentieth-century America. *Agricultural History, 65,* 1–12.

Daniel, P. (2013). *Dispossession: Discrimination against African American farmers in the age of civil rights.* Chapel Hill: UNC Press Books.

Davies, B. B., and Hodge, I. D. (2007). Exploring environmental perspectives in lowland agriculture: AQ methodology study in East Anglia, UK. *Ecological Economics, 61*(2–3), 323–333.

Du Bois, W. E. B. (1903). *The souls of black folk.* Chicago, IL: A. C. McClurg and Co.

Duhamel du Monceau, H.-L. (1762). *Eléments d'agriculture.* Paris: Guerin and Delatour.

F.B. (1672). *The office of the good housewife.* London: T. Ratcliffe and N. Thompson.

Fitzherbert, A. (1534). *The boke of husbandry.* Reprinted in W. Skeat (1882) "The book of husbandry." London: Trubner and Co.

Fussell, G. E. (1949). My impressions of William Marshall. *Agricultural History, 23*(1), 57–61.

Gerarde, J. (1597). *The Herball or Generall Historie of Plantes.* London: John Norton.

Glave, D. D. (2010). *Rooted in the Earth: Reclaiming the African American environmental heritage.* Chicago: Chicago Review Press.

Googe, B. (1578). *Fovre Bookes of husbandry.* London: John Wight.

Haggard, H. R. (1899). *A farmer's year being his commonplace book for 1893.* London: Longmans, Green, and Co.

Hersey, M. D. (2011). *My work is that of conservation: An environmental biography of George Washington Carver.* Athens and London: University of Georgia Press.

Hightower, J. (1978). *Hard tomatoes, hard times: The original Hightower report, unexpurgated, of the Agribusiness Accountability Project on the failure of America's land grant college complex and selected additional views of the problems and prospects of American agriculture in the late seventies.* Cambridge, MA: Schenkman Publishing.

Knapp, S. A. (1909). The farmer's cooperative demonstration work. *Yearbook of the Department of Agriculture 1909,* 153–160.

Lisle, E. (1757). *Observations in husbandry.* London: J. Hughs.

Loudon, J. (1839). *An encyclopaedia of agriculture.* Fourth edition. London: Longman, Orme, Brown, Green and Longmans.

Markham, G. (1616). *Cheape and Good Husbandry. For the well-ordering of all beasts, and Fowles, and for the general Cure of their Diseases.* London: T.S.

Markham, G. (1635). *The English husbandman.* London: Henry Taunton.

Marshall, W. (1778). *Minutes of agriculture, made on a farm of 200 acres of various soils.* London: J. Dodsley.

Martin, O. B. (1921). *The demonstration work: Dr. Seaman A. Knapp's contribution to civilization.* Boston, MA: Stratford Co.

Mavor, W. F. (1812). *Five hundred points of good husbandry: As well for the champion or open country, as for the woodland or several; together with A book of huswifery. Being a calendar of rural and domestic economy, for every month in the year; and exhibiting a picture of the agriculture, customs, and manners of England, in the sixteenth century.* London: Lackington, Allen and Co.

Mayberry, B. D. (1991). *A century of agriculture in the 1890 land-grant institutions and Tuskegee University, 1890–1990*. New York: Vantage Pres.

McDonald, D. (1908). *Agricultural writers from Sir Walter of Henly to Arthur Young. 1200–1800*. London: Horace Cox.

Merriam-Webster Dictionary (2018). www.merriam-webster.com/words-at-play/the-origin-of-farm (accessed 15 March 2020).

Payne, W., and Herrtage, S. J. (1878). *Thomas Tusser. Five hundred points of good husbandry. 1557, 1580*. London: Trübner.

Plat, H. (1594). *The jewel house of art and nature. Diverse new sorts of Soyle not yet bought into any Public use, for manuring both of pasture and arable ground*. London: Peter Short.

Ploeg, J. D. van der. (1994). Styles of farming: An introductory note on concepts and methodology. In J. D. van der Ploeg and A. Long (Eds.), *Born from within: Practice and perspectives of endogenous rural development* (pp. 2017–2227). Assen, The Netherlands: Van Gorcum.

Potter, C. (1998). *Against the grain: Agri-environmental reform in the United States and the European Union*. Wallingford: CAB International.

Primaudaye, P. d. l. (1589). *The French Academie, Wherein is discoursed the institution of manners*. Second edition. Trans. Thomas Bowes. London: Georg. Bishop.

Seavoy, R. E. (1998). *The American peasantry: Southern agricultural labor and its legacy, 1850–1995: A study in political economy*. Westport: Greenwood.

Skeat, W. W. (1882). *The book of husbandry by Master Fitzherbert*. London: Trübner and Co.

Smith, K. (2004). Black agrarianism and the foundations of black environmental thought. *Environmental Ethics, 26*, 267–286.

Surflet, R. (1600). *Estienne, Charles. Maison rustique or the Countrie Farme*. London: Edm. Bollifant.

Sutherland, L.-A., and Burton, R. J. F. (2011). Good farmers, good neighbours? The role of cultural capital in social capital development in a Scottish farming community. *Sociologia Ruralis, 51*(3), 238–255.

Sutherland, L.-A., and Darnhofer, I. (2012). Of organic farmers and 'good farmers': Changing habitus in rural England. *Journal of Rural Studies, 28*(3), 232–240.

Tusser, T. (1580). *Five hundred pointes of good husbandrie*. London: Henrie Denham. Reprinted in W. Payne and S. J. Herrtage (1878). London: Truebner and Co.

Von Münchhausen, O. (1766). *Der Hausvater. Zweiter Teil*. Hannover: Försters and Sons.

W. H. P. (1905). What is successful farming? *Michigan Farmer*, May 13, 469.

Waring, P. A. and Teller, W. M. (1944). The small farmer seeks a place in the Sun: He is sure the country needs him and that he can survive if he learns to cope with modern conditions and has proper support. The Small Farmer. *New York Times*, 12th March, p. SM18.

White, H. M. (1848). *Some of the five hundred points of good husbandry*. Oxford: John Henry Parker.

White, M. M. (2018). *Freedom farmers: Agricultural resistance and the black freedom movement: Justice, power, and politics*. Chapel Hill: University of North Carolina Press.

Worster, D. (1993). Good farming and the public good. In *The wealth of nature: Environmental history and the ecological imagination* (pp. 84–94). New York: Oxford University Press.

3

HOW SYMBOLS OF 'GOOD FARMING' DEVELOP

The historical development of 'tidy farming'

In Chapter 2, we illustrated how the concept of the good farmer developed and has been used in the literature since the emergence of the earliest agricultural publications. These broad notions of what a 'good farmer' was – based on God, family, community, production, stewardship, and so on – address the issue of what the role of the farmer is in society, community, and the family. However, as we outline in Chapter 1, the 'good farmer' concept is not a simple typological construct defined in the literature alone, but a living concept used by farming communities themselves to distinguish who is a 'good' farmer and who is not. Within these groups with local understandings of the practice of agriculture, the good farmer is constructed at a far more detailed level – on the basis of shared roles performed within the farming community and, in particular, the efficacy with which those roles are performed. As laid out in Burton's (1998, 2004a) seminal work on the issue, the 'good farmer' is constructed mainly from symbolic meanings associated with the performance of agricultural practices – where farmers are able to read from displays in the landscape the skills and knowledge applied to production. The good farmer's world is thus rich with symbols that represent both good farming and bad – visible to those with the experience and knowledge to understand them.

One of the criticisms that has emerged of the 'good farmer' is the notion that the concept is static or even deterministic. In particular, the focus of the first 'good farmer' research on productivist symbols and later outlining the difficulties of environmental symbols becoming part of the 'good farmer' (Burton, 2004a; Burton et al., 2008) has led some to warn against tying notions of good farming strictly to productivist behaviours – instead arguing that farmers' habitus is shifting (Riley, 2016a; Saunders, 2016; Sutherland and Darnhofer, 2012). This raises an interesting question. How do symbols of 'good farming' change?

A problem with studying symbolic change is the time frame over which such changes are likely to occur. The adoption of silage in the United Kingdom provides one example of this. Originating in continental Europe in the 1880s, it took a century before the practice – an ideal way to produce winter fodder in wet climates – was accepted as good farming in the United Kingdom despite being widely adopted in the United States a century earlier (Brassley, 1996). Numerous reasons are given for this including the weight relative to hay, the cost of the additional labour, the cost of building storage facilities, and even the smell of the product. Brassley also notes that in the first decade of the 20th century, the "cultural importance" of the then dominant root crop (the main alternative to silage) played a role in the lack of interest in silaging. Change towards silage began at the end of World War 1 as younger farmers began to take an interest in American methods. Today the production of silage dominates with 8.8 million tonnes of silage produced in the United Kingdom in 2017 – yet, it is the result of a debate that went on for more than a century and involved many generations of agriculturalists before reaching the final conclusion.

While it might be possible to gain an understanding of how symbols change by interviewing the farmers of today, using historical data allows us to explore change over an appropriate time frame without relying on recollections of distant memories. In this chapter we examine how maintaining tidy practices and a tidy landscape became important features of good farming (e.g. Birge and Herzon, 2019; Carr and Tait, 1991; von Glasenapp and Thornton, 2011; Guillem et al., 2012) over a period of decades. Maintaining a tidy landscape with modern agricultural equipment, skilled tractor-drivers and farm workers, and almost all aspects of the production process mechanised is relatively straightforward, but this was not always the case. Historically much of the farm work was done by unskilled labour by hand or with rudimentary instruments. As the 'tidy farm' is one of the most important symbols of good farming, we contend that understanding how the strong preference for tidy agriculture developed can provide us with the desired understanding of how symbolic meaning changes.

The value of untidy farming

The oldest guidebooks for practical farming in the English literature make little reference to farming in a tidy manner. Neither Fitzherbert (1534) nor Tusser (1580) emphasize neatness (such as the need to plough straight or turn over soil at a regular depth) as a requirement for good farming. Some mention of regularity in the landscape is made. For example, Tusser (1580) observes of hop farming:

> Five foot from another, each hillock would stand
> As straight as a levelled line with the hand
> Let every hillock be four feet wide,
> The better to come to, on every side.
>
> *(p. 98)*

However, here there is no extolling the virtues of tidy farming, only an observation on geometric patterns in agriculture. Without a substantial series of publications to draw from, it is difficult to establish with certainty whether tidy farming was part of the 'good farmer' identity in the past. Publications at the time were more concerned with outlining the basic practices of agriculture – such as the timing of plantings and other aspects of farm work – rather than describing whether they should be straight and tidy or not. Nevertheless, there is some indication that farmers were less concerned with tidiness. Rham (1845, p. 401), for example, observed how in earlier times farmers in Great Britain paid very little attention to the straightness of furrows – preferring to follow the crooked boundaries of the field. This, he notes, became such common practice that eventually "no straight furrow was to be seen; and there was a prejudice, if not a superstition, in favour of crooked ridges." In Lothian in Scotland, Buchan-Hepburn (1794) observed how it was only in 1723 that farmers John and Alexander Cunninghams began to level and replough ridges to straighten them. More importantly, he notes that the neighbouring farmers "at first derided the practice" but were later convinced by dramatic improvements in the crops such that the innovation quickly spread throughout the county. He notes that the benefit of having a clear and straight furrow

> struck every man so forcibly the instant he saw it, that the ridges in almost every corn field were straighted progressively as it came in its rotation to be fallow.
>
> *(p. 52)*

However, this practice does not seem to have spread rapidly everywhere. For example, it was only towards the end of the century that straight furrows "at last" became the preferred method of "every rational farmer" in Perth, Scotland (Robertson, 1794). James and Malcom (1794) similarly suggest that in the northern part of Buckinghamshire farmers "plough their lands in a complete serpentine form" which they suggested could be attributable to the "rich and fertile" soil meaning that farmers did not have to follow good farming practices.

Tidiness was not seen as necessary in a number of other situations. For example, a number of commentators complained that drawing straight boundaries in the process of enclosing the commons simply did not make any sense. Robertson (1793) referred to the straight field boundaries in Midlothian enclosures as "an absurdity" as they enclosed an array of different environments in a single field. Instead, he noted the field boundaries should vary depending both on the wetness of the land and the quality of the soil itself – such that "land of opposite qualities may not be included in the same close". In other places, customary beliefs dictated there was value to irregular landscapes. In Perthshire, Scotland, Robertson (1794) observed how boundary hedges in the region were invariably curved with "scarce ten yards of a straight hedge to be seen together in the county" which, he noted, was an extreme inconvenience to ploughmen. However, crooked hedges were believed to offer more shelter to cattle than straight hedges and thus were not straightened.

Buildings not maintained in a tidy fashion could also be seen in a positive light. For buildings, function was a primary concern, with the most attractive buildings being those that were "most convenient and useful to those who are concerned in them" (W.R., 1840, p. 373). This could mean maintaining buildings in an untidy form. An anonymous report in the *Quarterly Journal of Agriculture* (Anon, 1836, p. 332) observed that the much esteemed and valued Kerry (Ireland) butter was produced in "a shabby and apparently inconvenient house of mud walls, unplastered and without windows". This was not simply a matter of the financial position of the farmers. Rather, the shabby dairies were maintained in the belief that mud buildings without windows were best for production because "either lime or glass would be injurious to the butter process" while the high value of Kerry butter made it apparent that a well-constructed dairy house was "not absolutely necessary".

Untidy management of fields could also have benefits for farmers. One of the areas where this was important was in the use of non-agricultural plants ("weeds") as indicators of soil or climatic conditions that could benefit farm management. Many plants that grew incidentally around the fields alerted the farmer to underlying soil conditions, their potential for agricultural use and/or their need for improvement. For example, *The Complete Farmer* published by A Society of Gentlemen in 1777 discussed how valuable "the spontaneous produce of earth" was for assessing the nature and quality of farmland noting:

> Wild thyme shows it to be good for feeding cattle; betony and strawberries dried to wood; camomile points out a mellow soil fit for wheat; burnet indicates land fit for pasture; and mallows denote it proper for the uses of the kitchen-garden.
>
> *(p. 330)*

Throughout the 18th and early 19th centuries reference to the value of weeds as indicators is relatively common. The "gravel bind-weed" was a sign that gravel was lying near the surface (A Society of Gentlemen, 1777), "knob-weed" was seen as an indicator of poor quality land (Rudge, 1807), while "spurry" was a "sign of great poverty in a sandy soil" (Batchelor, 1808). Weeds could also indicate aspects of good farming. Pitt (1794) for example, observed that groundsel ("a garden weed") is plentiful in the Kentish hop grounds where it "is considered there as a sign of land in good condition". Although scientific knowledge of soils gradually replaced the importance of reading these signs in the fields, references to the importance of weeds were still being made into the second half of the 19th century. In 1862, Henry Cox (1862, p. 496) presented a paper to the Kingscote Agricultural Society on the natural history of weeds noting that "weeds too are of much use to the farmer in indicating to him the geological formation of the soil", and provided a long list of what weeds were believed to indicate about the soils they occupied.

In some places, farmers maintained their land in an apparently "untidy" or "slovenly" state because, in order to manage the farm, fields needed to produce more than simply agricultural crops. One example, of this is furze, which was noted by Potts (1809) as being tolerated on the uplands of Northamptonshire because of its value as fuel. In another case, farmers tolerated "scanty" oat crops as they believed it produced good straw for feeding to cows and horses – whereas if they grew 'good' oats, the stem would be too fibrous. The improver Sinclair (1814, p. 391) observed that "it is a miserable sign of the husbandry of any district, where such an observation is generally applicable". Yet farmers needed the straw to feed livestock.

One of the best examples of how necessity dictated good farming practices is that of the anthill meadows that were once relatively common across England (see Burton and Riley, 2018). Anthills were widely derided by improvers as a blight on the landscape because of the way they impinged on the ability of the farmer to implement effective meadow management – for example, making harvesting more difficult and leading to unusable wet areas. However, reports from the English county of Buckinghamshire around the turn of the 19th century suggested that many dairy farmers maintained large numbers of anthills on their meadows in the belief "with much seeming confidence" that anthills enhanced production due to the larger surface area on the field, produced an earlier flush of spring grass on the south side of the hills, and sheltered lambs in the spring (Potts, 1809 – also see Tusser, 1580, p. 59, who advises farmers to "make hillocks of mole-hills, in field throughout"). In the second report for Buckinghamshire, Priest (1810, p. 284) also observed that farmers in this region allowed anthills "in size scarcely to be credited" to form on the pastures such that "they are found not scattered here and there, but nearly as thick as the pasture can hold them from the tops of the ridges to the furrows". This, he noted, created an environment whereby water gathered in the unmown furrows and encouraged the growth of rushes. Priest viewed these as a "pernicious weed". However, when asking the farmers about them he noted

> more than one farmer informed me, that such rushes were valuable, and they should be very loth to exchange them for what others might esteem more productive plants, since such rushes served them instead of straw for their yards.
>
> *(p. 285)*

In this case, there was a clear discrepancy between the way improvers such as Priest and Young saw the anthills and how the farming communities involved viewed them. Despite this form of management being seen as an example of slovenly farming by the improvers, it took over a century after the first criticisms appeared in the literature in the early 1700s for ant-hill meadow management to disappear. Farmers' reluctance to abandon the practice (despite financial incentives from landlords) illustrates how farmers' views of what practices represent 'good farming' depended strongly on their needs. For the anthill meadows rushes were required in place of straw, for furze fuel was required because of a lack of alternatives, in Perth curved hedges were seen to provide necessary shelter, while the tolerance of scanty crops of oats was a compromise between the size of the oat crop and the need for palatable straw. In all these cases from the perspective of farmers the tidy farming symbols of neatly grassed hillsides, straight field boundaries, "weed" free properties, flat and easily managed hay meadows, and oat crops heavy with grain were diminished by the need to provide other essentials for the farm.

The development of tidy farming in fieldwork – furrows, ridges, and drilling

One of the earliest cases of agricultural authors promoting the virtues of tidy farming can be found in reference to the establishment of furrows. Straight furrows improved the drainage of the fields and enabled ploughing to be conducted with "no short turnings" – making it easier for less experienced ploughmen to plough the field straight. Symmetry was also an important part of the ridges. For optimal field management, ridges needed to not only be straight but parallel and "equally broad in all places" with the broadness and height selected depending on the soil type. The initial laying out of ridges in the field was also a skilled practice. Anon (1808a) observes how sometimes the furrow would not deviate over a 250–300 metre distance within a few centimetres from a straight line which he notes

> is not only a satisfactory evidence of the skill and dexterity of the ploughman in his art, but is really of importance towards the most perfect tillage of the land.
>
> *(p. 342)*

Badly constructed ridges created a number of problems. Firstly, if the ridge was not straight the plough would need to run in a curve, increasing physical resistance and making ploughing more strenuous. Secondly, in making ploughing more difficult, plough cuts often ended up being too wide or too narrow when the ploughman lost control. Thirdly, crooked furrows made sowing more difficult. Finally, irregular ridges could lead to horses causing "great harm" by treading on the crop when turning (Dickson, 1762; Kames, 1776). Tidiness in ridge construction thus became strongly connected to good farming practices.

Perhaps one of the most important practices in the development of the tidy-farming ethos was the invention of drill husbandry by Jethro Tull in the early part of the 18th century. Tull's invention began a slow transition from hand sowing to drill husbandry and, eventually, the abandonment of hand sowing entirely. Both practices required high levels of skill. However, the required skills were radically different and, consequently, so were the footprints they left on the farmed landscape.

Broadcast sowing involved a relatively simple act of scattering seed from a basket but was nevertheless a highly skilled job. The sower required a "measured step marking the time" in order to time the progress of the work, simultaneous actions of the limbs, a muscular body, and a "steady eye directing the flight of corn with a rainbow-like sweep to its destined place in the earth" (Anon, 1838, p. 104). Fitzherbert's (1534) description of sowing peas is a good example of how developing a rhythm was critical to being a good hand sower. He describes how the farmer should strap a bag filled with seed over his shoulder, and then

> set your left foot in front and take a handful of peas: and when you take up your right foot, then cast all the peas away; and when your left foot rises, take another handful, and when the right foot rises then cast them from you. And so at every two paces you will sow a handful of peas and if the foot and the hand agree, and then you will sow even. And in your casting, you must open your fingers as well as your hand, and the higher and further you cast your corn, the better it will spread.
>
> *(p. 19)*

The desired outcome was a thick and even coverage of seed. In order to ensure the dense coverage required to prevent weed growth among the crop (a key rationale behind broadcast sowing), large quantities of seed were often sown – making broadcast sowing an expensive process, and one where much of the seed sown was wasted. The lack of skilled hand sowers meant that the process was often done poorly by unskilled labourers.

Drilling, in contrast, involved geometric patterns, regularity, and tidiness. To drill seed properly the land needed to be properly prepared, the seed positioned in regularly organised trenches – all of an equal depth suited to the particular variety of seed being sown – and the seeds distributed at regular intervals. Rather than

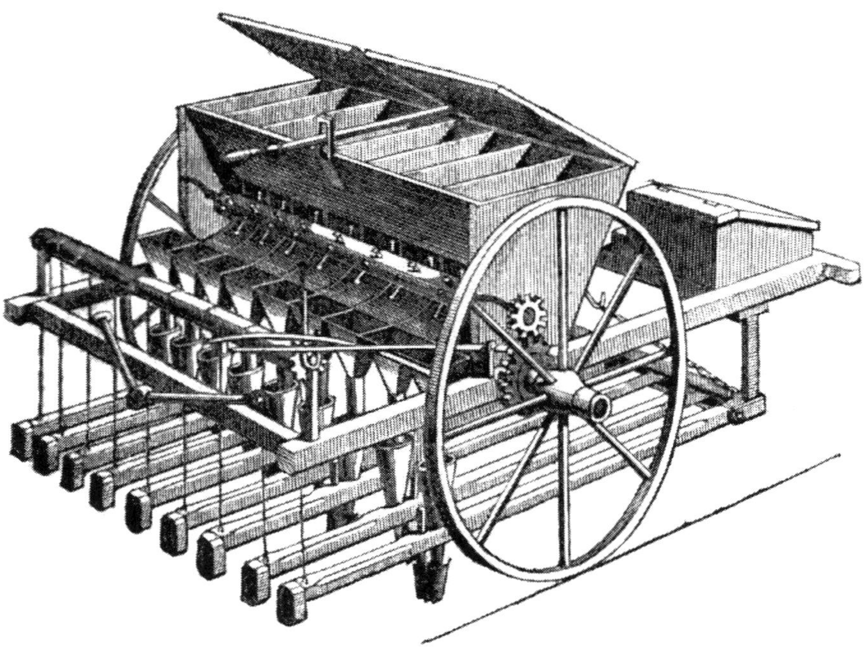

FIGURE 3.1 Norfolk Drill (Loudon, 1825, p. 387)

smothering weeds by blanketing the land with seed, drilling, by virtue of leaving gaps between the rows and planting crops in straight lines, allowed workers to get between the rows to hoe between the plants. Anon (1795) outlines the advantage of precision in planting:

> if the rows are placed very straight, (they ought to be exactly so) the shim will be of incomparable use, as it may be directed so near the rows as nearly to save hand work, except for extricating such weeds as grow among the plants, for which purpose a pronged hoe is of use.
>
> *(p. 95)*

This ability to use a hoe to remove weeds made drilling reportedly both more efficient and cheaper than hand sowing a crop (Anderdon, 1791). Hillyard (1837, p. 103) observed that the drill system was one of "the greatest improvements in the cultivation of land" primarily because it allowed the farmer to keep the land cleaner than was possible with broadcast sowing, with the ability to do so, he contended, being "the real difference between good and bad farming".

The move to drill husbandry also represented a considerable change in labour requirements. With broadcast sowing it was possible to take any labourer, even an unskilled one, and sow seed in a manner that would result in a crop. While skilled labour would do a better job, farmers were able to make do with what labour they had. Tools were simple. A basket was required to hold the seed, and, on emergence of the crop, hand tools could be deployed by labourers to remove the weeds. Preparation of the land for sowing could be as thorough or lax as the farmer desired, with some sowing done directly on unprepared fallow land.

Drill husbandry was an entirely different matter. To be drilled properly the soil needed to be carefully prepared, while the drilling itself was a precise operation – carried out with machinery that was often difficult to operate. Not only did operating the equipment require technical skills that were not widely available before the 19th century but repairing the machinery when it broke also required highly skilled technical labour. As a result, the success of drilled crops depended "upon the intelligence, attention, perseverance, and capital of the farmer" (Sinclair, 1821, p. 339).

Symbolically, drill husbandry also represented a significant change. With broadcast sowing, crops emerged as a surge of vegetation that (ideally) covered the field with an impenetrable blanket of green and signified suppression of any weeds, thus allowing the crop to grow without hindrance. Bare soil was an indication of failure where weeds were certain to establish themselves. With good drill husbandry, on the other hand, bare soil was essential for maintaining a weed-free crop and signified not the imminent arrival of weeds to compete with the crop, but that the crop could be readily managed with a hand or horse hoe. The pattern on the fields at emergence was thus one of straight drilled lines of thin crops with bare soil in between – complete anathema to those used to broadcast sowing.

The introduction of drill husbandry thus made interpreting signs of good farming difficult for farmers. After the initial emergence of the crop, where the gaps between rows were viewed as a sign of failure, the drilled crops grew strongly to the point where the productivity was greater than broadcast sown crops. However, there were still concerns that drilled crops were somehow fooling the farmer's eye. Agricultural improver Arthur Young (1792) suggests that drilled lucerne eventually began to look the same as broadcast sown lucerne but

> there is a fallacy to the eye in the drilled crops in proportion to the distance of the rows; they appear, thick while they are really thin, but in broadcast ones, which satisfy the eye, there is no deception, and these immense burthens, through which the scythe is with difficulty moved, produce more at one cutting than two-feet drills would at three with the advantage of the herbage being finer and softer.
>
> *(p. 325)*

Through generations of use, broadcast sowing had become part of the established culture of farming. Skills associated with broadcast sowing had come to represent a valuable commodity which "conveyed a superiority on the possessor" (Anon, 1838). The practice of sowing itself also had aesthetic qualities that led to strong sentiment developing about the importance of its presence in the countryside. The same observer notes:

> We own we should not like the idea of abandoning this ancient practice, though for nothing else than for the many pleasant associations by which it as endeared to every lover of field labour.
>
> *(p. 104)*

Many field experiments were carried out to test whether broadcast or drill husbandry produced higher yields, and these almost invariably came out in favour of drill husbandry. Nevertheless, hand sowing continued long into the 19th century. In some cases, the approaches were seen as being applicable in different circumstances. For example, broadcast sowing of lucerne was advised for situations where farmers were "doubtful of being able to give a very regular and constant attention to hand and horse hoeing" (Anon, 1796, p. 89). In other words, where farmers were unable or unwilling to pay the attention to the fields required for drill husbandry, broadcast sowing provided a workable alternative, and this is perhaps one reason broadcast husbandry was not rapidly overtaken by drill husbandry. Drill husbandry required not only new machinery, but a new type of farmer.

Ploughing – a critical symbol of good farming

The requirement of drill husbandry for well-prepared fields meant that skills in other areas of agriculture became increasingly important – in particular, those required for good ploughing. Ploughing and thereby ploughmen had long been recognised as socially important. Morey (1998, p. 41), for example, observes that ploughmen had a vital role in food production in the medieval era and, in recognition of their "obvious social value and their idealised moral status" ploughmen were provided with special legal protection, making the plough a form of sanctuary. On 19th century farms ploughing was equally valued, with ploughing cited as "the most important of all agricultural operations" as all subsequent operations depend on getting the ploughing correct (Anon, 1845, p. 63) and a "good ploughman" considered "the most necessary and useful man upon a farm" (Anon, 1843, p. 311). However, unlike drilling which involved revolutionary machinery, changes in ploughing – up until the invention of the "steam plough" in the mid-19th century at least – generally involved incremental improvements. By the mid-19th century, Coleman (1842) observed how the skill and art of ploughing had moved from digging the ground with a stick to its "perfect inversion" with a plough. As with drilling, ploughing had a geometric aesthetic quality of its own. He notes:

> A ruffle from under a crimping iron does not present a more beautiful object than a well-ploughed field from under the hands of an English or a Scotch artist. The lines are all straight; the furrows well turned; the headlands cross-ploughed; the corners finished. A well-disciplined mind enjoys the highest pleasure from seeing an operation of any kind, even the most humble, well performed, and perfected according to its proper measure.
>
> *(p. 461)*

The importance of ploughing to the good farmer is evident in the emergence of ploughing competitions. Formal matches appear to have begun with the Agricultural Societies of the 18th century, where premiums were offered in order to assist in the development of new technologies and to promote the improvement of agricultural practices (e.g. the Odiham Society of Agriculture and Industry, Baxter, 1785; Treby, 1792). In an article in the *Annals of Agriculture*, Erskine (1797, p. 332) asserts that the impact of these matches on ploughing skills in Clackmannan was substantial, with local ploughmen going from being "notorious for their want of skill in tillage" to "among the very best in Scotland" while now "the fields of the good farmers, indeed, appear cultivated like gardens". Reports of ploughing matches (which were frequent in England, Scotland, and Wales) suggest they played an important role in removing prejudices to new technologies, increasing the skills of ploughmen, and enhancing the geographical spread of innovation by enabling farmers to compare old technologies with new ones under similar conditions (Young, 1797; Sinclair, 1821).

Judging the skill of the ploughman in such a match was an expert task. Once again, straight lines were important. Clod (1844) observes that ploughmen in one match got into trouble at the start with veering furrows which he saw as problematic because

> there is no other part of his day's work, unless perhaps the finishing furrow, that demands so much expertness on the part of the ploughman as the veerings. It is requisite that they be straight and equal throughout.
>
> *(p. 520)*

However, this was not the only component of skilled ploughing. In a ploughing match at Glascuin organised by the Bishop of Kildare (Anon, 1808b) the judges described perfect ploughing:

> In all good ploughing of grass land for corn, we consider a perfectly straight furrow, cut perpendicular on the land side, at least five inches deep, the furrow cut completely horizontal at bottom, the sod of a moderate breadth, not thrown too much on the back, nor cocked too much on an edge, will be most likely to cover seed well, and to ensure a full crop.
>
> *(p. 307)*

The extent to which ploughing thus came to involve symmetry and regularity – the straightness of the lines, the angle the sods were cut at and the depth at which to which the plough was operating – illustrates the growing importance of regular landscapes and tidy fieldwork. Tidiness and regularity were seen as indicative of how well the seed would be covered and thus the ability of the work to "ensure a full crop". That this visible manifestation of good ploughing had social symbolic value can be seen in Clark's (1859, p. 210) observations of how ploughmen in Long Sutton did their "primest work" on Saturdays so that it could be viewed by passers-by going to church. This in turn, he notes, led to Sabbath-breaking where "groups of young men might be seen every Sunday taking their stroll round the district or through the marshes to see the straight furrows". The fact that farmers would risk the wrath of God (or their God-fearing neighbours) in order to view others' displays of good farming illustrates how important these social displays could be. It also illustrates how "roadside farming" (See Chapter 4 and Burton, 2004a) has long been an important means of transferring information on good farming credentials.

Constructing the 'good ploughman'

As with other forms of symbolic activity in agriculture, the visible work of the ploughman was simply a manifestation of an array of other skills necessary to be a good ploughman. Embodiment of the posture and instinct of a ploughman was one critical aspect – as noted by The Society for the Diffusion of Useful Knowledge (1837):

> There is, in fact, a certain degree of taste in ploughing as well as in everything else, a kind of tact which is difficult to be taught and hardly to be acquired, except by a sort of instinct, when once attained, it is perhaps the most healthy and agreeable of all agricultural exercises; the body being kept upright, the arms and legs brought into action, and also the eye and the mind to keep the furrow straight, and of regular width and depth, and the voice to speak to the horses.
>
> *(p. 41)*

Part of this embodiment was the development of the so-called voice – a way of talking that enabled the ploughmen to maintain control over the horses (Loudon, 1825). Clod (1844) observes how ploughmen often communicated with the horse through a series of "indiscriminately and indistinctly applied" sounds rather than issuing any clear directions to the horse. Repeated performance of ploughing also embedded skills at a subconscious level, for example, an understanding of how the plough moves through the soil (which enabled the ploughman to exert control over the plough), or being able to feel when the soil has been ploughed to a condition of "kindly mellowness" (Tucker et al., 1838).

In addition to these characteristics, many aspects of a good ploughman were related to personal qualities that were not developed through experience but were innate personality or physical characteristics of the individual. For example, Anon (1845) observed that brute strength (which, he contends, was the basis of ploughman selection for many farmers) is not all that is required to be a good ploughman, but he should be dexterous, show good judgement, keep an upright stance, be even tempered, and be attentive to the work.

Knowledge of a more conspicuous kind was also required – in particular setting the plough (and harness) was considered to be "one of the most difficult lessons of ploughmanship" (Johnson, 1842, p. 38). For farmers a store of experience-based knowledge on how to set the plough to suit particular crops, soil types and soil conditions (e.g. wet or dry) was a key part of being a good ploughman. For example, Boswell (1783, p. 361) observes how pillar-pins could be used in a Norfolk plough to raise, lower, and alter the direction and beam of the plough – thus setting it to plough deeper or shallower. A 'good ploughman' ploughing uneven ground would make these adjustments two or three times in a furrow.

At a time where ploughs were often locally manufactured, the construction and "goodness of the plough" could also affect the quality of the work (Young, 1797, p. 514). The type of plough was also important. A report on the "Implement department of the Warwick show, 1859" (R.S.B., 1860, p. 126) outlines a discussion on the "vexed question of the wheel and swing ploughs". Both types of plough had been in use for centuries by this time, with the wheel plough finally ousting the swing plough only with the arrival of the power tractor in the 20th century (Fussel, 1966). Each plough had its advantage. Using the wheel plough would "enable an inexperienced workman to perform work of a tolerably useful character, while the swing-plough would be almost entirely useless" to the same person who would have great difficulty ploughing to a uniform depth (R.S.B., 1860, p. 126). The author further suggested that the swing plough is an "implement" which requires considerable application of skill by the ploughman for correct operation but can respond to "the peculiarities of a soil or crop constantly varying in character". On the other hand, the wheel plough was seen as a "machine" which acts in the same way whatever soil it is acting upon. The machine was thus not able to do as good a job, but the ability to do it more economically made up for any loss in production. The skill displayed by the farmer and the tools used were thus inextricably intertwined.

FIGURE 3.2 Swing plough (Stephens, 1856, p. 3)

The preceding analysis of ploughing symbols suggests there are three main embodied components to producing symbols of good ploughing, namely:

1 Innate aspects of the body or person. While good ploughing could be learned, qualities associated with the basic physiology or personality of the farmer that contribute to being able to plough were also important in becoming a good ploughman. Some of these could be acquired through repeated practice in ploughing but others could not. The strong but dexterous, even-tempered, attentive farmer/farm worker was more likely to become a good ploughman than one lacking those qualities – no matter how much experience of ploughing he or she had.
2 Embodied characteristics learned through practice. Building on innate characteristics, the repeated practice of ploughing would lead to the development of embodied skills related to the way the ploughman connected to the physical world. This is effectively developing a "feel" for the practice of ploughing – whether it be through having the eye required to keep the furrow straight and regular, being able to hold a plough steady, having "the voice" to speak to the animals, the ability to read the body movements of animals to know instinctively how to respond, or the development of the correct body position required to be able to operate the plough. Ploughing patterned the ploughman.
3 Knowledge gained from practice or taught. A number of sources suggested that ploughing was a skill that could not be taught but only learned through experience. However, there were some aspects of ploughing that relied on learned knowledge: how to prepare the tackle for the horse, how to care for the horses, how to set a plough for the numerous different soil and weather conditions that were likely to be encountered, and even how to disassemble and reassemble a plough to move it from one field to another. Chomel (1725, p. 257) suggests the ploughman should also know the nature of the earth, the seasons of the year, and "the customs and fashions of the place wherein he lives". Unlike the other two categories that contribute to producing symbols of being a "good ploughman" this knowledge could be passed on directly.

In addition to these components of the 'good ploughman' the materials associated with the practice were also critical – in particular, the plough construction, the nature and condition of the animals, and the soil structure and condition. Unlike contemporary agriculture where much of the machinery has become standardised, historically ploughs were often designed and manufactured locally or even on the farm itself. Fitzherbert (1534), for example, observes

> There be ploughs of diverse makings in diverse counties, and in likewise there be ploughs of iron of diverse fashions. And that is because there may be many manner of grounds and soils. Some white clay, some red clay, some gravel or Chiltern, some sand, some normal earth, some mixed with marl, and in many places heath ground, and one plough will not serve in all places. Wherefore it is necessary to have diverse manners of ploughs.
>
> *(p. 2)*

This level of diversity was present well into the 1800s. As a result, the choice of plough (or the plough provided to the ploughman by the farmer) could make a significant difference in terms of the quality of the plough work. The training of the horses or oxen to walk at a steady pace also improved the quality of the performance (Lawson, 1827). Finally, the quality of the plough work was dependent on the material state of the soil – both in terms of its physical qualities (stony, peaty, clayey, etc.) and ephemeral condition caused by everyday variations in the weather. In this case knowledge of how to plough in different conditions and which equipment to use would affect the outcome.

Other straight lines and tidy symbols

While tidy sowing and ploughing were (and still are) key areas where good farming could be displayed, the general principle that untidy farms displayed poor farming practices extended across many other aspects of farm management. One area where tidiness was advocated was in the storage of harvested crops in stacks, cocks, and ricks – as these were both critical for successful farming and highly visible in the landscape. Oversized structures were advocated by farmers with limited practical experience in agriculture – "talk-workers" – but were strongly advised against by Lisle (1757) as they were unable to be pitched into a cart without being trampled and therefore shedding seed. The roofing of haystacks – which involved tying the hay tightly together to protect it against bad weather – allowed farmers to create ornate patterns that made the stack waterproof and provided the opportunity for the display of considerable skill. As the eaves of the stack were in particular danger of wind or rain damage, farmers were able to invest considerable effort in making them tidy without risking being accused of wasting time (Stephens and Norton, 1851).

Regular geometric patterns were an important part of indicating good practice in some cases. For example, Marshall (1778, p. 265) observed that for a stack frame a square (rather than a circle or parallelogram) is the best figure because it "is more pleasing to the eye, takes less thatch, and stands firmer, taken jointly, than on any other." It is interesting here that being "pleasing to the eye" is elevated in importance to lie alongside the practical advantages of the structure. In comparing bean stacks to oat stacks Marshall (1778, p. 264) asks, "which stack is the least handsome?" answering the bean stack because "the stem and the roof are out of proportion, and the bottom is too large". In an oat stack, having "too much belly" and a "heavy appearance" was a sign that the stack maker has been unable to keep the stems upright. The shape of stacks thus revealed aspects of the skill of their maker to those familiar with their construction.

Other highly visible constructions were also important symbols of good farming practices and, once again, the value of straight lines was recognised. Marshall (1778) for example, observes that

> In rural economy, straight lines and right angles are first principles, which can seldom be deviated from with propriety; either in laying out a farm, or in planning farm buildings.
>
> *(p. 125)*

However, it was not just straight lines that made farm structures symbolic of good practice. Other aspects of the construction also contributed. In particular, thatched straw rooves on buildings could reveal a great deal about the quality of the practices employed in harvesting and preparing the straw. Kent (1775) for example observed that:

> The common, injudicious, slovenly practice of beating the straw to pieces with the flail, and then laying it on with some of the seeds and many weeds in it, causes it very often to grow quite green, after it is laid upon the building; and, being bruised in all parts, to collect and retain the wet, much more than it would if the straw were whole, and consequently to become quite rotten in a few years.
>
> *(pp. 161–162)*

Thus, roofs thatched with straw prepared in an "injudicious" or "slovenly" manner would show clear signs of poor preparation – in the short term by sprouting green, and in the long term by rotting sooner than necessary. Kent also notes that when straw is meant for thatching, the corn should be cut earlier than usual. In contemporary cereal production cutting too early would carry an economic cost with no other benefit and would thus be read automatically as an indication of 'bad farming', but if a farmer were also concerned about straw quality the economic implications of early cutting would be less clear. This secondary use would make it more difficult to assess whether the early cutting was an indication of good or bad farming.

Fences too were required to be tidy. As with roadside farming in contemporary agriculture, the first thing passers-by would see from roadsides is the condition of the fences and hedges and, Anon (1869, p. 81) observed, "Broken fences and gates in a falling condition are a sure mark of slovenly management in other respects." In particular, ragged or wide hedges were seen by the improvers not as providing shelter for fields and livestock, but as occupying space that should have been used for productive farming (e.g. Bearn, 1852; Stephenson, 1874). Trimmed hedges, on the other hand, with the weeds pulled up from around the roots, were seen as good farming. However, work on the hedges – like other aspects of good farming – needed to be in proportion to their value in enhancing production as farmers could be criticised for expending too much labour on keeping the hedges tidy purely for aesthetic purposes (e.g. Stephenson, 1874).

Fluctuating symbols – economic conditions and control over land

Periods of boom and bust in agriculture driven by factors such as climatic abnormalities, changes in trading conditions, and the outbreak of war also affected the meaning of the symbolic displays. When it was considered unprofitable (or at least unnecessary) to be a tidy farmer, the value of tidy farming declined accordingly. Untidy farming was not looked on positively, but there was a level of understanding not present when conditions were good. For example, an observer writing in the *Farmer's Magazine* (Anon, 1835) observes that

> when good husbandry meets with so bad a return, the state of agriculture will retrograde, and a more slovenly system be substituted. This is even now the case in the hilly districts where improvement was making its way, but its march is absolutely stopped by the extreme distress of the farmers, who confine themselves to the supply of their own moderate wants from the produce of their lands.
>
> *(p. 135)*

Periods where agriculture was too profitable could also have a negative effect on good farming. A report in the *Farmer's Magazine* (Handley, 1838, p. 192) observed that during wartime any crop remunerated the grower and that "the most slovenly farmers amassed wealth, (as) the usual inducements to extraordinary exertions were wanting". Thus, historically, there were periods where 'good faming' was disincentivised by lack of return, while in other periods 'good farming' was disincentivised by the high profitability of even poor agricultural practices.

As well as the effect of economic boom and bust on good farming practices, some agricultural practices required significant amounts of capital that put 'good farming' outside the realm of the common farmer. Young (1771, p. 107), for example, notes of carrot production at the time that "such noble crops must not be expected without great expense" and that it is therefore clear that they cannot be grown by "common farmers, whose husbandry, in general, is so bad, from a want of money to carry on better".

The ability to produce symbols of good farming was also affected by the power of the landlords. Occasionally, farming customs or attempts to negate them became embedded in leases. For example, Lisle (1757) observes of the farmers in the "strong deep lands of Buckinghamshire" that prejudices against excessive barley planting had become institutionalised in leases that limited the planting of barley – even where it might have been good practice to do so. Likewise, Billingsly (1797, p. 303) noted that a tradition of over-exploiting land once it was ploughed – keeping it under constant tillage – led landlords to insist in the leases that no new land was to be tilled. The result was that the "regular and well-conceived rotation of crops . . . the most promising feature in good farming" was prevented for all farmers, good and bad.

The length of leases provided another impediment to good farming. Brown (1799) noted for Yorkshire, for example, that three-quarters of the land was rented from year to year, while even longer leases went for a maximum of only 21 years. This led farmers to think predominantly in the short term and limited farmers' custodial practices. The need to lengthen leases to improve agriculture was recognised and fought for by the agricultural improvers during the 19th century as, as Brown (1799, p. 30) noted, the renting of land by short-term lease which "to us seems destructive of good farming, is upon the increase". The end of the lease could also be problematic in cases where no compensation was paid for unexhausted improvements as it left no incentive for the farmer to steward the land properly for the next tenant (Caird, 1852).

Bad landlords could, and did, make good farming simply impossible. In explaining why observers of impoverished regions of agriculture with poor cultivation should not blame the farmers, a report of a meeting of the London Farmers' Club (Anon, 1848, p. 38) notes that underlying these practices was land that had not been drained, hedgerow trees that could not be cut down, dilapidated buildings, crops "overrun with game from the neighbouring preserves" and land that was "rack-rented" (rented at an exorbitant price). Trees maintained by landlords for timber in the hedgerows shaded the crops and limited the field sizes – thereby restricting the types of machinery that could gain access to the fields (Anon, 1848). The effect varied regionally. Caird (1852, p. 416), for example, observed of Northamptonshire that despite there being many excellent farmers "good farming was still the exception", with one of the key reasons being that landlords "have no interest in their farms beyond the annual rent they receive, know nothing of the management of land themselves, and do not employ an agent who does".

Lack of control over farming practices was clearly problematic for displaying symbols of 'good farming' historically. In particular, when developments in agricultural techniques and technologies began to gain pace in the 19th century, bad farming in some places was as much led by the failure of landlords to move from traditional leasing arrangements as by any failure of the farmers themselves. In this case, discerning the extent to which aspects of bad or good farming were attributable to the farmer would have been difficult. However, some symbols of good farming would have remained. For example, even where the farmer was required to use a cropping system considered by the peer group as bad farming, the skills involved in straight ploughing and other aspects of making a good job of field preparation would still be evident.

Today, subsidised production and high land values ensure that access to capital does not present the same limitations, rental agreements are generally not as prescriptive, and customary practices of landlords less prevalent than historically. In times of downturn, European farmers can fall back on subsidies for support without having to abandon all good farming practices. This said, subsidies and payments for environmental practices also allow new economic strategies oriented towards a maximisation of subsidies, even if this leads to practices that are contradictory with good farming symbols. At the same time, tenant farmers have greater legal rights while the power of the landlords has diminished considerably – leaving farmers free in most cases to make key decisions about farm management. As a result, the responsibility of the farmer for the appearance of his farm is stronger than may have been the case in the past. However, it is still sometimes possible to see when an invisible boundary between farms that are rented and farms that are owner occupied has been crossed, simply by observing the shabbier appearance of buildings and fence lines.

Putting it all together: the story of symbolic change in agriculture

Until the 19th century, tidiness did not dominate the ethos of the 'good farmer'. Seeds were often sown broadcast rather than being drilled in lines, and the primary objective was to smother as much of the land as possible with seed in order to prevent the emergence of weeds. Initially the pattern of straight lines with bare soil patches between was anathema to farmers. The debates that raged on the question of whether drill or broadcast husbandry was better practice – despite overwhelming evidence in favour of drill husbandry – illustrate the difficulties of moving away from more traditional symbols of good farming. Other 'untidy' practices could also be useful. In particular, a lack of soil science and geological knowledge meant that many of the non-agricultural plant species acted as indicators of the soil conditions and underlying geology, helping farmers to optimise their farm management practices. While weeds within field crops had no value, naturally occurring plants, "the spontaneous produce of the earth", in meadows and crop boundaries played an integral role in good farming. Knowing the conditions under which these indicator plants grew was of discernible economic value.

Prior to the widespread establishment of scientific agriculture, symbols of 'tidy farming' were somewhat difficult to interpret as the result of the multifunctional use of crops. In some cases, crops optimised for secondary purposes were not organised and tidy in the way food crops were. Whether tolerating poor oat crops (as measured in grain yield) in order to provide better straw for feeding to cows and horses, harvesting crops early for straw, growing furze for fuel, or allowing ant hills to develop to provide rushes for use as straw, the lack of necessary materials for other aspects of farm management meant that "tidy" production did not always serve the farmer's needs.

This gradually changed. Increased knowledge of plant physiology, better understanding of soil chemistry and geology, and the availability of geological mapping progressively reduced the need for tolerating non-crop plants in the fields. At the same time, transport systems in the 18th century had been notoriously bad – limiting both the export of agricultural produce and the import of other necessities such as materials for roofing, heating, or fodder for animals. As part of the wider industrialisation processes, improved roading systems and the development of canals and railways opened up areas of the countryside to the import and export of goods. As such the need for secondary uses of crops diminished. Pressure from the 'agricultural improvers' of the time was also heavily directed towards optimising production in market-based agriculture – to the point where actions that failed to optimise productivity, even when there was a practical purpose for doing so, were widely derided (see Burton and Riley, 2018).

A key factor in the move towards 'tidy farming' was the development of improved agricultural machinery and management techniques. Whether in terms of the invention of ploughs that enabled less-skilled workers to plough straighter lines to a more even depth, or the development of drills that dispersed seeds in straight rows and at even depth, almost every step towards mechanisation was a step towards tidiness and regularity becoming an important symbol of good farming practice. Even the development of a broadcast sowing machine led to more regular patterns in the landscape, with the scattering of seed no longer dependent on the ability of the sower to walk at an even pace and move limbs with uniform rhythm, potentially resulting in bare patches or over-seeding.

Simultaneously, the development of new machinery began to erode the skills necessary for 'good farming'. Whereas simple agricultural implements required operation by workers with innate skills, practice-based skills, and knowledge-based skills, agricultural machinery made these embodied aspects of 'good farming' increasingly redundant. No longer was a skilled ploughman (who demanded high wages) required to plough a difficult field with a swing plough. Instead, a relatively unskilled labourer could operate the wheeled plough. Although not as efficient in terms of being able to respond to changes in soil conditions as the swing plough, the lower cost of labour meant that, overall, it was a more economic means of production and, on average, resulted in higher production as unskilled labour performing skilled jobs led to poor outcomes.

The move from broadcast sowing to drill sowing was one of the most critical moments. Studies comparing horse drill and horse hoe agriculture to conventional broadcast agriculture repeatedly illustrated that crop yields were higher under drill agriculture. However, traditionally the symbol of a good farmer had been a lush, verdant carpet of green emerging after broadcast sowing. Farmers struggled both with the concept that drill husbandry left large patches of open ground that weeds could colonise (as the main form of weed control had been to outcompete the weeds with crop plants) and that the thin lines of plants resulting from drill husbandry could possibly produce more than the thick carpets of green evident in broadcast sowing. Eventually a generational shift occurred as farmers who were neither attached to the old ways nor ignorant of the value of the scientific evidence took over.

The development of the 'tidy farm' ethos presented in this chapter has provided some interesting lessons concerning the long-term development of symbols of good farming.

First, technological advances in agriculture have a critical role in the development of 'tidy farming symbols. Machinery leaves an imprint on the landscape that is distinctive, visible, regular, and often symmetrical. Our current array of agricultural implements pulled by powerful tractors can produce highly regular forms in almost all soil types – restricting symbols of poor performance to factors such as working the soil when it is too wet, setting the machinery incorrectly, or failing to set straight tramlines when sowing the crop (see Burton, 2012). For other farming practices, technology removed all symbolic importance by making the practice redundant. For example, while the tidiness with which stacks and ricks were finished was once an important symbol of good farming, round baling and silage production have eliminated the need for drying or field storage of hay in most cases – and thus the construction of weather resistant structures in the fields. As technologies and practices change, new symbols emerge and others diminish and disappear.

Second, changes to what symbolises good farming can take decades or even generations. The length of time required for a symbol to lose or gain meaning is related to a number of factors including the strength of customary beliefs (attachment to old practices), the availability of the technology or the skills required, and the extent to which the practice/technology is considered an improvement over previous approaches. In the move from broadcast to drill husbandry, complex mechanical engineering in the drills limited farmers' ability to purchase, operate, and maintain the drills, while the straight lines and open soil appearance was simply too different to the pattern produced by broadcast sowing for it to be easily accepted.

Third, as the wider socio-economic and environmental context within which farming takes place changes, so can the symbols of good farming. Examples such as managing ant-hill meadows for reeds for thatch or producing furze on the hillsides for fuel show how what is considered 'good farming' is tailored to the farming community's needs outside as well as inside agricultural production. The demise of ant-hill meadows was in part the result of the arrival of alternative sources of straw, while leaving furze on the hillsides is likely to have been locally considered a symbol of bad farming once coal could be obtained through improved transport links.

Fourth, the analysis of the 'good ploughman' provides us with a framework for identifying the components of the 'good farmer'. A good ploughman is a combination of innate features of personality and physiology, embodied characteristics that have been incorporated through practice (a 'feel' for the plough), and the knowledge gained through either experience or teaching. The extent to which the work of the ploughman is symbolic of 'good ploughing' is then moderated by material factors such as the training of the animals, the equipment available, the nature of the soil, or the control exerted over the practice by the landlord. We contend this can also be applied as a general model of the qualities of a good farmer – to display symbols of good farming the farmer needs the necessary innate, embodied, and knowledge characteristics with the performance moderated by the tools and materials at his/her disposal.

Perhaps the most important observation is that there appears to be a progressive decline in the symbolic wealth of agricultural landscape over the centuries. Historically, farmed landscapes were more complex arenas of practice with beliefs about good farming locally embedded, machinery locally constructed for local conditions, shortages of farming materials creating local needs for dual purposes in the cropping, a need for ecological as well as agricultural knowledge, and so on. The countryside of today is much more symbolically standardised. Because of the globalisation of food production (in particular crop varieties and machinery), a farmer from Scotland can read and understand the significance of cropping displays from a farmer in the south of England, continental Europe, or even further afield with relative ease. It is this homogeneity that has contributed to 'tidy farming' being globally recognised as important as a symbol across conventional industrial farming cultures (Burton, 2012). In the future, as systems are increasingly globalised and new technologies remove the risk of sub-optimal farming competencies, we expect the number of symbols that represent displays of 'good farming' will continue to decline. Ultimately smart or precision farming could eliminate symbolic displays of competence from the land entirely and replace them with hidden skills such as knowledge of machinery maintenance, IT skills, and specific knowledge of relevant computer software. Symbols that have had a significant role in farming communities for centuries – if not millennia – could become relics of the past.

References

A Society of Gentlemen (1777). *The complete farmer.* Third edition. Printed for J. F, and C. Rivinoton, J. F., London: Rivinoton, C., Crowder, S., Longman, T., Law, B., Lowndes, T., Robinson, G., Cadell, T., and R. Baldwin.

Anderdon, R. P. (1791). Account of a crop of turnips, raised between rows of beans, hand-hoed and horse-hoed. With observations on drilling and horse-hoeing Turnips. *Letters and Papers on Agriculture, Planting, Etc. addressed to the Society instituted at Bath, 3*, 117–132.

Anon (1796). Some notes concerning lucern. *Annals of Agriculture and Other Useful Arts, 25*, 85–107.

Anon (1808a). Essex tillage. *Annals of Agriculture and Other Useful Arts, 45*, 340–355.

Anon (1808b). Farming society of Ireland: Dublin Spring Shew (sic). *The Agricultural Magazine, 2*, 305–313.

Anon (1835). Agricultural reports. Glamorgan. *The Farmer's Magazine NS 2*, 135–136.

Anon (1836). On the agriculture of the county of Kerry. *The Quarterly Journal of Agriculture, 7*, 317–344.

Anon (1838). The agriculturalist's notebook no. 3. *The Quarterly Journal of Agriculture, 7*, 64–118.

Anon (1843). Agricultural report for September. *Agricultural Journal and Transactions of the Lower Canada Agricultural Society, 1*(10), 308–313.

Anon (1845). Essay on ploughing. *The Agricultural Magazine, 1*, 63–64.

Anon (1848). What are the practical impediments to the development of British husbandry? *The Farmer's Magazine, S2*(17), 34–51.

Anon (1869). Calendar of agriculture. *The Farmer's Magazine, S3*(36), 80–81.

Batchelor, T. (1808). *General view of the agriculture of the county of Bedford*. London: McMillan.

Baxter, A. (1785). Transactions of the Odiham society of agriculture and industry. *Annals of Agriculture and Other Useful Arts, 3*, 304–310.

Bearn, W. (1852). Farming of Northamptonshire. *Journal of the Royal Agricultural Society, 13*, 44–113.

Billingsly, J. (1797). *General view of the agriculture of the county of Somerset*. London: R. Cuttwell.

Birge, T., and Herzon, I. (2019). Exploring cultural acceptability of a hypothetical results-based agri-environment payment for grassland biodiversity. *Journal of Rural Studies, 67*, 1–11.

Boswell, G. (1783). Description of the model of a Norfolk plough sent to the Society by Mr Boswell. *Letters and Papers on Agriculture, Planting, Etc. addressed to the Society instituted at Bath, 2*, 356–361.

Brassley, P. (1996). Silage in Britain, 1880–1990: The delayed adoption of an innovation. *The Agricultural History Review, 44*(1), 63–87.

Brown, R. (1799). *General view of the agriculture of the county of Somerset*. Edinburgh: James Watson.

Buchan-Hepburn, G. (1794). *General view of the agriculture and rural economy of East Lothian*. Edinburgh: John Moir.

Burton, R. J. F. (1998). *The role of farmer self-identity in agricultural decision making in the Marston Vale Community Forest*. Unpublished PhD Thesis. De Montfort University, Leicester.

Burton, R. J. F. (2004). Seeing through the 'good farmer's' eyes: Towards developing an understanding of the social symbolic value of 'productivist' behaviour. *Sociologia Ruralis, 44*(2), 195–215.

Burton, R. J. F. (2012). Understanding farmers' aesthetic preference for tidy agricultural landscapes: A bourdieusian perspective. *Landscape Research, 37*(1), 51–71.

Burton, R. J. F., Kuczera, C., and Schwarz, G. (2008). Exploring farmers' cultural resistance to voluntary agri-environmental schemes. *Sociologia Ruralis, 48*(1), 16–37.

Burton, R. J. F., and Riley, M. (2018). Traditional ecological knowledge from the internet? The case of hay meadows in Europe. *Land Use Policy, 70*, 334–346.

Caird, J. (1852). *English agriculture in 1850–51*. Second edition. London: Longman, Brown, Green and Longmans.

Carr, S., and Tait, J. (1991). Differences in the attitudes of farmers and conservationists and their implications. *Journal of Environmental Management, 32*, 281–294.

Chomel, M. (1725). *Dictionary Oeconomique*. Vol. 2. London: D. Midwinter.

Clark, J. (1859). A farmer's life and duties. *The Farmer's Magazine, S3*(15), 210–212.

Clod (1844). On ploughing. *The Farmer's Magazine, S2*(9), 520–521.

Coleman, H. (1842). *European agriculture and rural economy from personal observation*. Vol. 1. Boston: Joseph Rogerson.

Cox, H. (1862). On the natural history of weeds, and the best means of extirpating them. *The Farmer's Magazine, S3*(21), 496–500.

Dickson, A. (1762). *A treatise of agriculture*. Vol. 1. A. London: Donaldson and J. Reid.

Erskine, F. (1797). Manufacture of glass in Clackmanan. *Annals of Agriculture and Other Useful Arts*, *25*, 218–221.

Fitzherbert, A. (1534). *The boke of husbandry*. Reprinted in W. Skeat (1882) "The book of husbandry". London: Trubner & Co.

Fussell, G. E. (1966). Ploughs and ploughing before 1800. *Agricultural History*, *40*(3), 177–186.

Guillem, E. E., Barnes, A. P., Rounsevell, M. D. A., and Renwick, A. (2012). Refining perception-based farmer typologies with the analysis of past census data. *Journal of Environmental Management*, *110*, 226–235.

Handley, H. (1838). Improvement in agriculture. National agricultural institution. *The Farmer's Magazine*, *NS*(8), 191–198.

Hillyard, C. (1837). *Practical farming and grazing*. Second edition. Northampton: T.E. Dicey.

James, W., and Malcom, J. (1794). *General view of the agriculture of the county of Buckingham*. London: Colin Macrae.

Johnson, C. W. (1842). *British husbandry exhibiting the farming practice in various parts of the United Kingdom*. Vol. II. London: Robert Baldwin.

Kames, H. (1776). *The gentleman farmer*. Edinburgh: W. Creech.

Kent, N. (1775). *Hints to gentlemen of landed property*. London: J. Dodsley.

Lawson, A. (1827). *The farmer's practical instructor*. London: Mackenzie and Dent.

Lisle, E. (1757). *Observations in husbandry*. London: J. Hughs.

Loudon, J. C. (1825). *An encyclopaedia of agriculture*. London: Longman, Hurst, Rees, Orme, Brown, and Green.

Marshall, W. (1778). *Minutes of agriculture, made on a farm of 200 acres of various soils*. London: J. Dodsley.

Morey, J. H. (1998). Plows, laws, and sanctuary in medieval England and in the wakefield "Mactacio Abel". *Studies in Philology*, *95*(1), 41–55.

Pitt, W. (1794). *General view of the agriculture of the county of Stafford*. London: T. Wright.

Potts, T. (1809). *The British farmer's cyclopaedia being a new and complete agricultural dictionary of improved modern husbandry*. Second edition. London: B. Crosby.

Priest, J. (1810). *General view of the agriculture of Buckinghamshire*. London: Richard Phillips.

R. S. B. (1860). A retrospect of the leading features of the implement department of the Warwick show, 1859. *The Farmer's Magazine*, *S3*(18), 125–128.

Rham, W. L. (1845). *The dictionary of the farm*. Second edition. London: Charles Knight and Co.

Riley, M. (2016a). How does longer term participation in agri-environment schemes [re] shape farmers' environmental dispositions and identities? *Land Use Policy*, *52*, 62–75.

Robertson, G. (1793). *General view of the agriculture of the county of Midlothian*. J. Edinburgh: Anderson.

Robertson, G. (1794). *General view of the agriculture in the Southern districts of the county of Perth*. London: J. Nichols.

Rudge, T. (1807). *General view of the agriculture of the county of Gloucester*. London: Richard Phillips.

Saunders, F. P. (2016). Complex shades of green: Gradually changing notions of the 'Good Farmer' in a Swedish context. *Sociologia Ruralis*, *56*(3), 391–407.

Sinclair, J. (1814). *An account of the systems of husbandry adopted in the more improved districts of Scotland*. Vol. 1. Third edition. Edinburgh: John Brown.

Sinclair, J. (1821). *The code of agriculture*. Fifth edition. Edinburgh: Sherwood, Gilbert & Piper, and Tait.

Stephens, H. (1856). *Catechism of practical agriculture*. Edinburgh and London: William Blackwood and Sons.

Stephens, H., and Norton, J. P. (1851). *The farmer's guide to scientific and practical agriculture*. Vol. 2. New York: Leonard Scott and Co.

Stephenson, H. (1874). Report on the farm prize competition, 1873. *Journal of the Royal Agricultural Society, S2*(10), 246–264.

Sutherland, L.-A., and Darnhofer, I. (2012). Of organic farmers and 'good farmers': Changing habitus in rural England. *Journal of Rural Studies, 28*(3), 232–240.

The Society for the Diffusion of Useful Knowledge (1837). *British husbandry; Exhibiting the farming practice in various parts of the United Kingdom*. Vol. 2. London: Baldwin and Cradock.

Treby, P. T. (1792). *Rules, orders and premiums of the South Devon Division of the Devonshire Society for the Encouragement of Agriculture and Industry*. Plymouth: B. Haydon.

Tucker, L., Gaylord, W., and Thomas, J. J. (1838). *The Genesee farmer and gardener's journal*. Vol. 8. Rochestire: Luther Tucker.

Tusser, T. (1580). *Five hundred pointes of good Husbandrie*. London: Henrie Denham. Reprinted in W. Payne and S. J. Herrtage (1878). London: Truebner & Co.

Von Glasenapp, M., and Thornton, T. F. (2011). Traditional ecological knowledge of Swiss alpine farmers and their resilience to socioecological change. *Human Ecology, 39*, 769–781.

W. R. (1840). Science for Farmers. *The New England Farmer and Horticultural Register, 18*(45), 373.

Young, A. (1771). *The farmer's tour through the East of England*. Vol. 4. London: W. Strahan and W. Nill.

Young, A. (1792). *Travels during the years 1787, 1788, and 1789. Undertaken more practically with a view of ascertaining the cultivation, wealth, resources and national prosperity of the Kingdom of France*. Vol. 2. Second edition. London: J. Rackham.

Young, A. (1797). Prizes for ploughing with oxen at Petworth. *Annals of Agriculture and Other Useful Arts, 29*, 511–520.

4

THEORISING THE 'GOOD FARMER'

From common sense category to analytical construct

In the historical literature, the term 'good farmer' was employed in a multitude of ways. Early authors such as Tusser and Fitzherbert sought to develop a moral peasantry, agriculturalists used it to validate their ideas by citing 'good farmers' as approving (where there was no experimental evidence), while improvers used the concept to dismiss customs by seeking to synonymise the term with productive commercial agriculture. Ultimately, it was the improvers who took control of the term. Throughout the 18th and 19th centuries a relentless narrative supporting modernisation sought to move agriculture from a peasant way of life to an industry. While the 'good farmer' could be 'good' at caring for God, the family and community, or 'good' at farming in a customary manner where productive tidy agriculture was not the priority, increasingly the 'good farmer' became he/she who fought superstition and tradition to produce as much as possible out of the land – for queen, country, and for the sake of progress itself. This narrative of the 'good farmer' as a good producer accompanied the professionalisation of agriculture, where the 'farmer' progressively moved away from a definition encompassing the whole personhood, including explicit references to morality and the place and role in family and community, to a dimension related mainly to the mastering of specialised skills and techniques in the name of production. Through this process, moral and cultural aspects of modern farming have been hidden under a veil of scientific rationality.

As a consequence much of the literature on the 'good farmer' provides us with 'official' perspectives on good farming – be it the agricultural improvers from the 18th and 19th centuries, the demonstration farm developers in the United States around the turn of the 20th century, or the policy makers of the post–World War II productivist era in the mid-20th century. These production-oriented perspectives have been motivated by a wide range of factors such as food insecurity, wars, population growth, a desire for economic development, and so on. However, while increasing production was important from the perspective of those in power, it does not necessarily represent the primary objective of the farmers themselves. Many of the 'common' farmers those days were not involved in formally defining the 'good farmer', as most were illiterate. For example, Verral (1799) observing of farmers in his region:

> Books treating scientifically on agriculture are of little, if any utility, to the small wealdish [referring to the Weald, UK] farmer. For want of education, his capacity, expanded only in a very limited degree, is not capable of understanding the improving lessons genius exhibits.
>
> *(p. 100)*

Thus, the historical literature analysis provides us with an indication of the kind of 'good farmer' the elites wanted – with a very normative tone aimed at transforming contemporary agriculture and very few texts written by practical farmers themselves. While in Chapter 3 we pieced together an understanding of how symbols of the 'good famer' change from historical literature, this is of limited assistance in understanding other aspects of the good farmer – such as the role of cultural capital in the generation of social capital (e.g. Sutherland and Burton, 2012), how concepts of the 'good farmer' affect farmer identities, and how the desire to be a 'good farmer' contributes to the everyday practice of farming. From the perspective of contemporary research, the good farmer concept is important precisely because it offers these kinds of insights into the farming culture. The understanding of how symbols of good farming develop and change over time is a significant step towards being able to utilise the 'good farmer' concept as an analytical tool; however, we also need to understand how the concept fits in with contemporary social theory.

This leads us to a key issue for the book. Numerous researchers have employed the concept of the 'good farmer' as a theoretical framework – and have done so in a variety of ways. However, as noted in Chapter 1, the 'good farmer' is not a theory. Unlike Mead's/Goffman's symbolic interactionism or Bourdieu's theory of the economy of practices, for the 'good farmer' there is no conceptual ownership, no body of literature outlining, defending, and theoretically developing the concept – nor is there any single theoretical framework around which the 'good farmer' literature is based. Frameworks are selected generally on the basis they might assist the researcher to understand the particular aspect of behaviour under investigation, whether it be why productivist behaviours are symbolically meaningful to farmers (Burton, 2004a), why conventional farmers are often resistant to agri-environmental policy (Burton et al., 2008), how farmer identity affects conversion to organic farming (Sutherland and Darnhofer, 2012; Sutherland, 2013) or adaptation to a multifunctionality regime (Forney, 2010), how farming cultural values change under market liberalisation (Haggerty et al., 2009), or any of the many other topics relating to the cultures of agriculture.

Thus, in order to be able to operationalise the good farmer in research, an understanding of the theoretical frameworks underlying the use of the term is essential. To this end, here we present the main frameworks that have been used with the concept of the good farmer, an analysis of their various strengths and weaknesses, and a discussion concerning how the concept has evolved from what we call a 'common sense category' to an 'analytical construct'.

The good farmer – from common sense category to analytical construct?

In its most basic understanding, the term 'good farmer' simply reflects how the observer regards the observed – 's/he is a good farmer' or 's/he is a bad farmer' – and is the language necessary to express a preference. It is an external judgement, made by more or less qualified observers; a 'common sense category' that represents an evaluation of competency (as described in Chapter 1). This common sense category approach is employed specifically to understand the issue from the actor's perspective.

Here we view the term 'good farmer' primarily as a 'positional label' – reflecting identification with a particular group and determined by the performance of roles that are culturally viewed as defining group belonging (see Stryker, 1980). For example, in order to be classified as a farmer, one might need to 'produce crops', 'manage livestock', 'keep the farm tidy', 'get your boots dirty', 'demonstrate independence', 'keep the name on the land', and so on. In this sense, the term 'good farmer' emerges because being able to self-identify as a farmer is not only a matter of performing the role, but performing the role to the level of competency expected by others within the social group. For example, as a result of bad management, livestock can be injured, crops can fail, and farms can be 'tidy' to the point that production roles are compromised. To be recognised as a 'farmer' each behavioural role must be performed well – reward for which is peer group recognition as a 'good farmer'. The 'good farmer' is therefore, in this simplest sense, a 'farmer' who has reached the necessary level of competency.

Recognition, however, also depends on which farming behaviours, beliefs, values or symbols are important in identification, and for much 'good farmer' research the focus of investigation is on the conventional productivist farmer. However, while a focus on production continues to dominate farming communities (see Walford, 2003; Mather et al., 2006; Wilson and Burton, 2015), farmers comprise many different cultural groups and sub-groups – each with their own set of 'good farming' practices. Factors such as local environmental conditions, the targeting of different sectors of the market (e.g. organic farming), or the geographical isolation of the community promote the emergence of distinct definitions of the 'good farmer' (further discussed in Chapter 7). This is also evident in the 'farming styles' work of Ploeg and others (e.g. Howden and Vanclay, 2000; Ploeg, 1994, 2003; Vanclay et al., 2006) as well as research into how the 'good farmer' is constructed from the perspective of organic farmers (Stock, 2007; Sutherland and Darnhofer, 2012; Sutherland, 2013). While these definitions share common elements – in particular the production role which remains the core of the farmer identity – in other areas there can be important differences. For example, the prioritisation of conservation, profit achievement, non-farming roles, or traditional production approaches can serve to define the farmer as a 'good farmer' within his or her peer group, but may exclude their membership of a different peer group with alternative visions of farming (see Burton and Wilson, 2006).

Understanding the symbolism behind these behaviours moves the concept of the 'good farmer' away from a simple categorically descriptive term towards presenting a meaningful cultural construction defined by a cultural repertoire of normative and strategic ideas about what constitutes good practice. In a sense, the researcher moves from focusing on the 'good farmer' with unspecified roles (for example, mentions of "he is a good farmer" in interview texts), to focusing investigation on the roles themselves and their cultural construction in order to define specific identity groups and explore contrasts, conflicts, alterations, etc. in the repertoire. In this sense, the 'good farmer' becomes an analytical category underlain by and embedded within a cultural system, and it is in understanding the significance of being recognised as a good farmer at both the individual and community level that its utility as a research tool emerges. Examples include how good farmer credentials lead to the development of social capital within the farming community and thus place a social value on agricultural behaviours (e.g. Sutherland and Burton, 2011), how transitioning to part-time farming may lead to loss of 'good farmer' identity (Sutherland, 2019), and how the importance of being visually recognised as a good farmer affects the 'tidy farm' ethos (e.g. Burton et al., 2008; Burton, 2012). In this way, the 'good farmer' acts as a focus for understanding the cultural importance of farming activities and therefore farming culture itself.

The concept of the 'good farmer' as an analytical category has been used by numerous authors seeking to understand how farmers' self-conceptions and cultural influences influence their behaviour within the wider context of agrarian change. In the next section of this chapter we review this literature to illustrate the different uses of the good farmer concept. In particular, as the good farmer is an analytical rather than a theoretical concept, we explore how researchers have borrowed theoretical perspectives in order to conceptualise and operationalise the cultural norms underlying the 'good farmer'.

Symbolic interactionist interpretations of the 'good farmer'

For symbolic interactionists, the individual and society are characterised as part of a dynamic, constantly interacting system in which the self is conceptualised as "essentially a social structure [that] . . . arises from its social experiences" (Mead, 1934, p. 140). Mead argues that "our thinking always takes place by means of some sort of symbols" (p. 146), and that, as the meaning of these symbols has been negotiated through interaction with society, every action and object has a meaning of shared significance to both parties. Self-identity develops as the individual interacts with the social group and learns the group meanings. Through such interactions meaning is being constantly renegotiated. Eventually, social structures such as language, interpretative procedures, attitudes, roles, and social class perspectives become internalised (Coughenour, 1976; Weigert et al., 1986) – i.e. the individual begins to view them as his/her own – and adopts the 'self-referent label' or 'positional label' (Stryker, 1980) of the group, e.g. 'I am a *farmer*'. Membership of the group in the eyes of others is developed and maintained through displaying commitment to the same symbolic meanings as held by the wider group by performing the appropriate role behaviours 'Farmers do . . .' In this way, every individual is a reflection of his/her socio-cultural upbringing.

Two conceptualisations of the good farmer employ a symbolic interactionist perspective, namely Silvasti (2003a, 2003b) and Burton (Burton, 1998, 2004a, Burton and Wilson, 2006). In the first of these cases, Silvasti (2003a) applied the notion of "social scripting" (also termed "cultural scripting" – Vanclay and Enticott, 2011) to establishing a cultural model of "the good farmer" in Finland. Social scripting is "a logical extension of symbolic interactionism" – based on the work of Blumer and Mead, but often associated with Goffman and the metaphor of the theatrical production (Wiederman, 2015, p. 10). In outlining the core theoretical framework for her approach, Silvasti contends that

> Social interactionism [*sic*] defines "social scripting" as a process, where persons are subconsciously and consciously conditioned to follow rules, and adapt values and behavioural patterns determined by society, its subculture, or some ethnic or socioeconomic group.
>
> *(p. 144)*

As is standard with symbolic interactionism, this process is seen as a matter of the socialisation of cultural meanings, instilling beliefs about appropriate role-behaviours and collective symbolic understandings. According to Silvasti, these are termed "scripts", and their internalisation allows people to organise behaviours along socially appropriate lines – including behaviours related to 'good farming'. As with many other interpretations of the 'good farmer', Silvasti's use of the 'good farmer' concept within the social scripting framework is based around the notion of farm work and, in particular, what is seen as "real work" – working the fields, caring for farm animals, and maintaining the buildings. Symbolically, she notes, the appearance of the farm is seen by others within the community as representative of the "kind of farmer that farms there" with farmers gaining considerable intrinsic satisfaction from the concrete results of their actions. Social scripts create strong emotional ties between the family and the farm and influence the way of life of the farm family.

The notion of scripting as outlined by Silvasti has been applied by other researchers in studies of farming culture. For example, Rogers et al. (2013, p. 255) refer to social scripting as "farmer talk . . . such as what it means to be a good farmer" with, again, an emphasis on socialisation as a means of embedding 'rules' that farmers subconsciously follow. Again following Silvasti, Herman (2015) contends that normalised social scripts define what behaviours are acceptable if one is to be seen as a 'good farmer' and argues that these normalised social scripts determining what defines the 'good farmer' provide a boundary framework within which social resilience takes place. Finally, in perhaps the most theoretically elaborated piece, Vanclay and Enticott (2011) argue that scripts represent a "story of common line of argument" that provides farmers with a rationale or justification for a particular course of action. The authors identify a number of different types of script that exist in rural communities, namely, scripts as expected routines, scripts as catch phrases, scripts as parables or short stories, and scripts as commonly used lines of argument. These scripts provide group members with a sense of "the right thing to do" that can be applied in different contexts.

The second use of a symbolic interactionist framework connected to the 'good farmer' literature can be found in the early work of Burton (Burton, 1998, 2004a; Burton and Wilson, 2006). Whereas the social scripting approach draws from sociologist Erving Goffman's interpretation of Mead (Vanclay and Enticott, 2011), Burton's draws from social psychologist Sheldon Stryker's social structuralist "identity theory". Identity theory was developed to clarify some of the ideas of Mead (1934) and translate them into terms with clearer empirical refinement (Stryker, 1994). In particular, the theory focuses on two aspects connected to behaviour, namely the salience of an identity within a hierarchy of identities (where identity is seen as multiple, structured, and hierarchical), and the degree of commitment expressed by the individual to the roles that define the identity group. Identity theory deals only with role behaviours where the individual has a choice between alternative courses of action and sees the selection of a behaviour as determined by a combination of the salience of the identity and the commitment the individual has to the social group. Commitment is measured as the loss of relationships with significant others (measured both in terms of affective loss and quantitative loss of interactions) were the person to choose a course of action that falls outside the remit of the identity group (Stryker, 1968, 1987) – in our case, the roles of the 'good farmer' (see Burton, 1998; Burton and Wilson, 2006 for further details on Stryker's theory). In choosing a behaviour, the individual thus weighs up – often subconsciously – how much s/he stands to lose or gain from the salient identity group if s/he chooses to undertake the behaviour.

Burton's (2004a) work, "Seeing through the 'good farmer's' eyes: towards developing an understanding of the social symbolic value of 'productivist' behaviour", was arguably the first publication to emerge using this theoretical framework. This paper does not employ identity theory in its entirety, but rather uses the theory to justify the exploration of the symbolic value of production activities to explore how the "farmer" identity is constructed – i.e. the symbolic meanings of the role performances associated with the 'good farmer'. While Goffman's "dramaturgical metaphor" is mentioned in the paper, the paper does not follow the "social scripting" approach of Silvasti (2003a, 2003b). Rather, it explores the meanings behind practical farming behaviour and, importantly, how symbols of good farming are identified, generated, and transmitted – a concept Burton later investigates using the theories of Pierre Bourdieu (see section that follows).

Nevertheless, the symbolic interactionist theory of Stryker has been applied. Burton and Wilson (2006), in particular, explore the connection between the salience of known farmer identities (commonly typologised groups) with behaviour on the farm. The authors conclude that complex and mobile identity structures are one explanation for farmers being able to see themselves as conservation-oriented while carrying out seemingly contrary behaviours. In other words, despite concerns that farmers' behaviour is still overwhelmingly productivist, nature conservation role-behaviours can be part of the 'good farmer' identity, but these may not always be salient. McGuire et al. (2013) used this framework to develop a "Good farmer identity control model" to explain how a farmer's identity could shift from being predominantly productivist to predominantly conservationist. This was used to distinguish between farmers with conservationist identities and productivist identities and explore how this affected their willingness to undertake water quality improvements. The authors concluded that through engagement with water quality measures the position of the "conservationist" identity moved higher up in the identity hierarchy – with the result that roles associated with 'good farming' were gradually changing. The multiple identity approach as applied by Burton and Wilson (2006), while offering an interesting perspective on farmer identity, is difficult to operationalise and perhaps is unsuited as a framework for looking at the 'good farmer', relying, as it does, on a rather rigorous quantitative methodology.

Bourdieusian approaches

While interest in symbolic interactionist approaches to the 'good farmer' appears to have been limited, the theories of French sociologist Pierre Bourdieu are perhaps the most widespread theories used in 'good farmer' analyses (e.g. Burton et al., 2008; Burton and Paragahawewa, 2011; Droz, 2002; Forney, 2007; Glover, 2010; Hunt, 2010; Raedeke et al., 2003; Riley, 2016a, 2016b; Riley et al., 2018; Sutherland and Burton, 2011; Saunders, 2016; Shortall et al., 2018; Sutherland and Darnhofer, 2012; Sutherland, 2013, 2019). Fundamentally Bourdieu's (1998) theory contains three components – habitus, field, and capital. In brief, Bourdieu defines the habitus (and field) as

> a socialised body, a structured body, a body which has incorporated the immanent structures of a world or of a particular sector of that world – a field – and which structures the perception of that world as well as action within that world.
>
> *(p. 81)*

Habitus, he contends, is the centre for the formulation of "generative schemes" which guide action – predominantly because the individual has the sense that it is the right thing to do rather than as the result of conscious reasoning. In an often-quoted phrase, it provides a "feel for the game" or practical sense (Bourdieu,

1990, p 66). The concept of capital is thoroughly outlined in Bourdieu's (1986) paper which details a "general science of the economy of practices" (p. 243). Here Bourdieu postulates that our focus on economic capital is due to the unambiguous immediacy of economic exchange – whereas capital can be stored in the other more ambiguous forms of social and cultural capital. Capital can be transformed from one form to another via symbolic capital. Symbolic capital is no specific form of capital but rather a condition whereby those with a shared habitus are able to recognise the significance of the capital through having developed a shared understanding of its symbolic value because "for a symbolic exchange to function, the two parties must have identical categories of perception and appreciation" (Bourdieu, 1998, p. 100).[1]

Pierre Bourdieu's (2008) early work in the Bearn dealt with the transformation of the rural society in the 1960s, when the peasants' habitus became a disadvantage to individuals on the matrimonial "market". He himself did not engage in any discussion on the cultural and social construction of the 'good farmer'. One of the first references to Bourdieu in the context of the 'good farmer' can be found in Shucksmith's (1993, p. 468) paper on farm household transition to post-productivism. Here Shucksmith contends that for farmers, the habitus derives in large part from the "subconscious and cumulative assimilation of an established ethos of being a farmer". In another early paper, Tsouvalis et al. (2000, p. 912) employ habitus to conceptualise farmer knowledge cultures as "socially negotiated structures of meaning that enable and constrain social actions" which they contend enable us to "acknowledge the fluid and interactive nature of different ways of sense-making" and highlight "the formative contexts within which meaningful, symbolic actions and knowledges are shaped." Raedeke et al. (2003) use the notion of habitus to explain internalised subconscious directives for action that provide farmers with a feel for the game. In their analysis of agroforestry adoption, they recognise the economic importance of being seen within a community as a 'good farmer' in that landlords are more likely to rent out farms to those who maintain a good farming ethos.

A Bourdieusian framework was also employed in an ethnographic study of a farming community in the Swiss Jura in the late 1990s by anthropologist Yvan Droz (Droz and Miéville-Ott, 2001; Droz, 2002). Droz and colleagues' approach was based on another Bourdieusian concept, that of "ethos". Ethos, as defined by Droz and Lavigne (2006), is a non-structured set of guiding principles for action that are both laden with moral values and symbols and generally acknowledged as legitimate and just. As part of the habitus, the ethos is constructed, embodied, and reproduced through a socialisation process that includes direct experiences and learning processes (Droz and Lavigne, 2006, p. 48). The embodied nature of the ethos differentiates it from ethics or morality: the ethos is anchored as much in small habits, everyday practices, bodily postures, as it is in explicitly moral or normative narratives.

Haggerty et al. (2009) advanced the notion of habitus to explore the deregulation and intensification of sheep farming in New Zealand. Here the authors examine how notions of good farming with respect to animal health change with the intensification of agriculture beginning with the Bourdieusian premise that farmer responses to structural changes in the broader "field" are generated in part by the relationship between habitus and field, and suggest farmers have acquired their "feel for the game" through long-term participation in agriculture. They note:

> If Bourdieu is correct, and the pursuit of social status is fundamental to social life, farmers will strive to be "good farmers" not "bad farmers", to accrue "good farmer" capital, and this drive will inform their farming strategies.
>
> *(p. 769)*

The strength of Bourdieu's theory, according to Haggerty and colleagues, lies in its ability to explain social reproduction and resistance to change, and more concretely in the case of the New Zealand sheep farmers "that it helps makes sense of the lack of the development of environmental subjectivities" (p. 769). However, beyond narratives of conservatism, it places these resistances in a context of social struggles between differently positioned actors, as well as "between strategies, dispositions, and thinkable and unthinkable courses of actions" (p. 769). Haggerty and colleagues further note that the cultural capital associated with competency in stockmanship has become a core part of the 'good farmer' identity amongst New Zealand sheep farmers, and that market intensification has forced farmers to abandon some of these values in order to maintain profitability. As with the previous studies focusing on habitus, this paper leaves the issue of how practices become part of the 'good farmer' ethos largely unaddressed.

The third way Bourdieusian theory has been employed is to use systems of capital rather than habitus as the key to understanding good farming. The first application of this approach was Burton et al.'s (2008) study into farmers' adoption of agri-environmental behaviours. Burton and colleagues adopted Bourdieu's systems of capital as a framework because of its acknowledgement of the non-economic symbolic importance of farming behaviours – strongly indicated by earlier findings on the social symbolic value of productivist behaviours (Burton, 2004b). The notion that farming behaviours could be of cultural and social as well as economic value provided a framework for understanding why farmers might 'sacrifice' profit (for example, by spending uneconomic amounts of money on machinery or keeping roadsides tidy) in order to be seen as a good farmer by others or reject agri-environmental schemes because of the loss of status as a 'good farmer'.

In the 2008 study, Burton and colleagues explored how the "identical categories of perception and appreciation" noted by Bourdieu as critical for the establishment of symbolic capital develop through the everyday practice of agriculture. Three qualities required for behaviours to become symbols of good farming are postulated. First, the outcome of the application of cultural capital should reflect the extent of the capital held by the farmer (predominantly in the form of skills and knowledge). An action where skill/knowledge has little impact on the outcome is unlikely to become of symbolic importance. Second, the outcome of the application of cultural capital needs to be visibly manifest as opposed to hidden. Financial returns and yields that are known only to the farmer him/herself are less important as symbols of good farming than those where the farmer is unable to hide the outcome (e.g. the production of a tidy crop or a well-maintained property) (Burton, 2004a). Third, other farmers must have access to the visible displays of cultural competency. This is generally through observation from the roadsides and is the basis of the notion of roadside framing (see Chapter 1) but may also be through more formal learning processes such as farm discussion groups.

Later papers developed these ideas further. For example, Burton and Paragahawewa (2011) explore how agri-environmental schemes might be made more culturally sustainable, Sutherland and Burton (2011) explore how cultural capital can be converted into social capital, and Burton (2012) examines how the cultural importance of visual displays of farming affect farmers' aesthetic appreciation of landscape. Others have also focused on the importance of capital to the 'good farmer' concept. Butler and Holloway (2015), for example, hypothesise on the role of "hybrid-capitals" for redefining what is culturally valuable as 'good farming' in farming systems where rapid technological changes are altering the relationships between farmers and livestock. Conway et al. (2016) use the concept of symbolic capital as a framework for understanding how older farmers prioritise the maintenance of non-economic capital over ceasing agricultural activity – even where the activities are economically unviable. Riley (2016, p. 96) similarly uses "Bourdieusian-inspired discussions of good farming" to explore the issue of how symbols of good farming change as the farmer ages.

Sutherland and Darnhofer (2012) and Sutherland (2013) focus more specifically on the processes underlying the formation and alteration of good farming identity. Their papers demonstrate specific conditions under which symbols change, drawing on an English dataset of organic, conventional, and farmers in conversion. In line with Burton, the papers draw on Bourdieusian concepts of capital exchange: economic capital (material and financial property), social capital (networks of social connections and mutual obligations), and cultural capital (prestige, status in the community). In Sutherland and Darnhofer (2012) the conceptual focus is on Bourdieu's concepts 'habitus', 'field', and 'rules of the game'. Following Bourdieu (1998), the habitus is conceptualised as an unconscious disposition to act, developed through interactions within specific 'fields' – arenas of production characterised by actors who compete to accumulate different forms of capital (Swartz, 1997, p. 117). By engaging in these fields (e.g. agricultural commodity markets), the 'rules of the game' become internalised in the participants' habitus. When the rules of the game change, creative responses are required. Change may also occur without changing rules, through creative action (e.g. Jenkins, 1992), but the focus in the good farming literature has largely been on resilience.

Sutherland and Darnhofer (2012) identify several processes that lead to change in the symbols of good farming: devaluation of current symbols, reflexivity, and experimentation (further elaborated in Chapter 7). The unprofitability of 'good farming practices' devalues associated symbols, leading to an openness to consider other farming options, and enter different 'fields', with associated new 'rules of the game'. Through reflexive consideration of new practices, new skill development, and rewards for new practices (e.g. higher prices for organic produce, positive feedback from customers), current symbols are renegotiated and sometimes replaced. Sutherland (2013) took these arguments further by introducing Bourdieu's concept of the 'taste of necessity'. Bourdieu (1984) argues that a taste for luxury is evident when cultural capital (e.g. evident in works of contemporary art) has limited connection to the production of economic capital. Luxury is evident in appreciating and owning objects with little practical utility. In contrast, a 'taste of necessity' is reflected in a direct economic relationship between cultural symbols and economic limitations (e.g. a preference for beer and cheap, high calorie food among lower socio-economic classes) (Bourdieu, 1984, p. 173). In essence, virtue is made of necessity.

Sutherland (2013) argues that the symbols characterising cultural capital in contemporary farming largely reflect necessity – weed-free fields represent higher productivity and thus greater household income; healthy animals are more likely to reproduce and gain maximum value at auction sales (also see Burton, 2012). Although the cultural value of particular symbols (pedigree livestock, for example) may exceed their apparent commercial value, the link between economic and cultural capital remains readily apparent. 'Good farmers' are successful farmers, who maintain a profitable farming enterprise. However, it was also clear that this connection, while direct, is not the only issue – farmers seek ideals to aspire to beyond the financial. Their discourse is overtly negative when farmers are seen as solely profit oriented. 'Good farming' thus lies on a continuum between necessity and luxury. Formerly commercial farmers who opted to continue farming on a recreational or hobby basis thus identified their actions as a step towards luxury – the opportunity to engage in a valued activity without need for commercial gain.

Particular aspects of rule changes are reflected in the speed of symbolic change. The ease of change is important – incremental shifts are more easily implemented. For example, keeping up with technology leads to new symbols with limited resistance – the objectified cultural capital of a new piece of farm machinery devalues over time, as newer models are introduced. The consistency of these rules is very important – inconsistent signals in any of the fields in which farmers engage make it difficult to form new symbols: the direct connection between symbols and outcomes is not evident. In the Sutherland and Darnhofer cases studied, once it became evident that new policies (e.g. environmental regulations) and opportunities (such as agri-environmental schemes) were long term, they became embedded in the rules of the game. The fluctuating value of organic farming commodity prices was a different story – organic conversion represented both a major change to farming trajectory and an uncertain outcome – subsidies oriented towards enabling organic farming conversion initiated shortly before the study had led to a market glut and resultant drop in commodity prices. The potential for new symbols to be visually perceived and evaluated, in line with 'roadside farming', is also a factor – new practices that have no visual symbol (or appear negative, such as the weedy fields characteristic of organic farming) are particularly problematic. Key in organic farm conversion was the potential to develop new symbols, such as varied fields, and visual appreciation of environmental actions such as beetle banks.

While many others have addressed the good farmer concept using a Bourdieusian framework, Mark Riley and colleagues' work represents probably the most significant single body of research. Riley's work appears in a number of key papers. First, Riley (2016a) explored how farmers' participation in agri-environmental schemes shapes environmental identities. In this paper, Riley introduces the notion of hysteresis into the good farmer literature. Hysteresis, it is contended, occurs when there is discord between the habitus and the field and causes a structural lag between the opportunities available and the dispositions to grasp them (Bourdieu, 1990). Riley uses the concept of hysteresis to explore changes in symbolic meaning, in particular, how definitions of the good farmer vary spatially and how the good farmer concept evolves over time. In a second paper, Riley (2016b) used the good farmer notion as a conceptual framework to explore how farmers maintain farming identities even after retirement from the farm. Again looking at the evolution of meaning, the study found that the symbolic significance of farming activities changes during the life course of the farmer – thus enabling semi-retired farmers to maintain their self-identity as a 'good farmer' even when they are no longer performing key symbolic roles. In a third paper, Riley et al. (2018) focusing again on agri-environmental measures using a Bourdieusian framework, observe how collective agri-environmental scheme participation plays out in a social arena where maintaining reputation as a good farmer plays a critical role. Collective environmental behaviours are thus moderated by concerns for the maintenance of social and cultural capital. Riley and colleagues have also looked at alternative communities and productions following the good farmer concept, in particular Gustavsson et al.'s (2017) work looking at the development of social capital and networks of reciprocity in fishing communities, and Thomas et al.'s (2019) exploration of riparian environments which found that farmers engagement with rivers is limited by their inability to generate cultural capital.

Other theoretical engagements with the 'good farmer' analytical construct

While the symbolic interactionist and Bourdieusian approaches dominate as theoretical frameworks behind the good farmer construct, other approaches to theorising the 'good farmer' exist – some of which make only minor reference to the concept, while others present promising frameworks for future exploration. One example is Dessein and Neven's (2007) framing of "farmer pride" drawing on the social and economic psychology literature on self-esteem. As with other 'good farmer' applications the authors focus on multiple aspects of identity that comprise personal and collective identity, postulating that it is the interaction between the personal and collective identities in a constantly changing context that results in "farmer pride".

Forney (2010), having initially contributed to the Bourdieusian theorisation of the 'good farmer' through the concept of ethos (as stated previously), proposed a theorisation based on the concept of social representation as defined by the French social psychologists Moscovici, Abric, and Jodelet (Abric, 1994; Jodelet, 2007). The purpose was to better include social interactions and a dynamic and evolving dimension in the conceptualisation of farmers' identities. Social representations are "a kind of knowledge socially constructed and shared, with a practical aim, and participating to the common construction of reality within a social group" (Jodelet, 2007, p. 36).[2] The social representation framing stays close to the ethos perspective developed by Droz et al. (see earlier); however, it introduces nuance and flexibility in what is otherwise a slightly rigid conceptualisation. Rather than assembling a complete array of values covering diverse facets of farmers' identities, social representations simply frame the narratives and practices that form diverse and sometimes competing images (figures) of what is or should be a farmer. Social representations also focus not on individuals and how their inclusion in a group influences its behaviour – the *farmer's* ethos – but rather on the object of the representation and on the interactions that create it, i.e. representations *of the profession*.

Using this framework, Forney identified four different representations (figures) of the good farmer: food producer, succeeded successor, the independent worker, and the manager. Those figures do not correspond to four categories of farmers; rather they summarise four general patterns of this self-representation that are transversally acknowledged as legitimate, even if individuals might not identify as strongly with each of them. Drawing on the concept of posture developed with Droz and colleagues to describe landscape representations (Droz et al., 2009), he suggests that every farmer articulates in a personal and varying way these four representations. This framing allows the integration of diversity and individual interpretations while acknowledging a shared identification with a 'good farmer' ideal.

The theory of the core of social representations that differentiate between peripheral and central elements of the representations (Abric, 1994) is useful for understanding the evolving nature of the 'good farmer'. Drawing on Flament's (Flament, 2007) development on the dynamism of representations, Forney suggests that peripheral elements evolve in time and vary from one person to the other, without affecting the core of the farmers' self-representations. This explains how self-representations adapt to external changes at their periphery (e.g. a good production move from quantitative criteria to qualitative ones), while remaining fundamentally untouched at the core (the importance of producing food). More fundamental changes happen progressively, step-by-step, or after a major shock due to radical changes in the context.

The concept of social representation also allows the role played by non-farmers to be included in the construction of the 'good farmer'. Representations are constructed in social interactions between farmers, but also with other actors (farmers' organisations, the administration, industries, citizens, etc.). As an illustration of this co-production of self-representations, the four figures constituting the 'good farmer' ideal in the early 21st century Swiss farming milieu clearly echo some aspects of the normative discourses on the farmers developed by agrarian political elites in the early 20th century, who played an influential role in the construction of post-war productivist agricultural policies (Forney, 2011). This illustrates the role of elite discourses in the shaping of the 'good farmer' narrative as outlined in Chapter 2. As a relatively recent introduction to the good farmer literature, conceptualisations based on the notion of social representation have yet to emerge in the wider literature; however, the approach is broadly applied in a wide array of fields (e.g. renewable energy, health and disease).

Huttunen and Peltoma (2016) integrate Bourdieusian inspired 'good farming' concepts with practice theory, making the connection through Bourdieu's 'logic of practices'. The power of 'practice' is in its performance, which is situated in time and space. Practice theory thus draws attention to the materiality of good farming symbols. Consistent with Sutherland and Darnhofer (2012), Huttunen and Peltoma argue that through farmers engaging in environmental practices, the meaning and practices associated with long-held ideals (e.g. of farmers as environmental stewards, farming as productive) are renegotiated, and new symbols are identified and reinforced. New symbols are more likely to be adopted if they represent a progression, rather running counter, to existing farming practices and symbols.

Finally, Emery (2010) engages in a sophisticated theoretical discussion on farming cultures and the 'good farmer'. Inspired by the rhetoric-culture approach by Carrithers (2005a, 2005b), his theorisation starts from a definition of culture as dynamic. Culture is made and re-made by actors who use it strategically and negotiate it continuously. Three aspects are central here: the interaction; the persuasion of oneself or of others; and the creation of new cultural forms through reinterpretation and re-making of existing elements. In this general discussion on culture, Emery (2010) insists on the propensity for individuals to be both agent and patient, to both act on and be acted upon. While all people are capable of strategically and creatively using cultural elements in order to adapt to changing situations or to pursue their own interests, social interactions are embedded in structural power relations. In this rhetorical approach to culture, power is understood first as the capacity to use effectively the 'means of persuasion', and second, in a more pervasive way, as the ability to determine what those means of persuasion will be.

Applying this to the discussion on the 'good farmer' and changing farming cultures, this means that changes in values reflect a continued negotiation/renegotiation between farmers and others, where individual farmers use and negotiate strategically this repertoire of values to build their identity – or 'personhood' – but acknowledge that this repertoire includes also structures of social power. Here, identification with those values is not necessarily related to a social competition for prestige or domination. Emery recalls Burke's (1969) argument on the necessity of identification (and its accompanying concepts of particularisation and differentiation) as constitutive to the facts of being human and of communicating. The need to identify and be identified is the major driving force for engaging with farming values. From this perspective, personhood is only momentarily constructed through "rhetorical attempts in the play of agents and patients and contingent upon their setting or *situation*" (p. 22), as identification is a matter of continuous efforts of rhetorical persuasion. However, the rhetoric culture approach should not be misinterpreted as a reduction of identity to discourses. The values expressed by farmers in their processes of identification are tied to practice as much as to narratives and are embodied in the landscape, in the farm work, and in the livestock.

The rhetoric culture approach to farming cultures emphasizes the agency of the farmers in the use and negotiation of symbols of the 'good farmer' and, therefore, the flexibility and diversity of possible interpretations and enactments of what is a good farmer. Farming values are seen as 'things of possibility', capable of being interpreted and used in many different ways and situations. Therefore, there is never one definition of the 'good farmer', but only momentary attempts at convincing others and oneself that someone is a good farmer.

Theorisation of the good farmer: a discussion

This diverse range of theories underlying the 'good farmer' construct all contribute to a movement of the term away from a normative and common-sense construct towards a more theoretical – and from our perspective useful – conceptualisation. The differences in focuses and framing given previously define the explanatory potential of a specific theorisation, by highlighting (or neglecting) specific processes in the social and cultural production of farming cultures. Moreover, the 'good farmer' analytical framing is often referred to in more or less applied research endeavours. Consequently, its theorisation will impact on how scientists, extensionists, rural developers, or any other actors using this framing, address the question of the cultural dimension of farming, and what answers to addressing 'cultural resistance' they might propose. It is therefore important to discuss, in brief, the pros and cons of the different theorisations.

Silvasti's (2003a, 2003b) use of symbolic interactionism makes an important contribution of the theorisation of the good farmer by highlighting the importance of socialisation and cultural transmission, by referring to social scripting as "a process, where individuals are subconsciously and consciously conditioned to follow rules, and adapt values and behavioural patterns determined by society, its subculture, or some ethnic or socioeconomic group" (p. 144). The processes of socialisation, notably by the relatives and the social milieu, and the role of the belonging to a group are undoubtedly important facets of the cultural dimension of farming. However, in her discussion of Thompson's "three religious-philosophical tenets behind productionism" (p. 144), her analysis remains largely attached to discourses and narratives and does not engage with practices and social interactions as important aspects of culture. Although there are clearly a broad set of socially transmitted beliefs concerning the need for tidy farming, maintaining the name on the land, and so on (as identified by Silvasti, 2003a and many others), we contend that role-behaviours associated with good farming (i.e. the set of behaviours and values that can be used to identify a 'good farmer') develop mostly through the *practice of agriculture* as farmers learn what it means to be able to farm well and how to evaluate the 'good farmer' credentials of others. Similarly, her conceptualisation of "script" does not provide any conceptual means of understanding how the values and symbols of the good farmer are generated and transformed through time.

Some applications of the Bourdieusian concept of habitus to the good farmer concept present a similar emphasis on social belonging and reproduction. When Droz (2002, 2001) and colleagues (Droz and Forney, 2007; Droz and Miéville-Ott, 2001) describe the "peasant ethos", the focus is in the embodiment of values that are transmitted and shared within the farming community and contested by changes in public policies. Because of their insistence on the strength and importance of the cultural dimension of farming, there is a tendency to enact a slightly monolithic and static definition of culture. The confrontation with the new logics carried by a changing agricultural policy is consequently interpreted as a cultural and identity clash, resulting from a process of harsh symbolic and political domination.

Both Silvasti and Droz and colleagues produce an image of a farming culture "under threat" (Silvasti's wording). Interestingly, the two approaches nevertheless diverge in their fundamental positioning. While Silvasti's aim is to understand why Finnish farmers do not accept a critical environmental reading of their practices, Droz criticises the pressures on the Swiss farmers' identity resulting from a neo-liberal turn in agricultural policies. These diverging approaches, however, both concentrate on resistances produced by the social reproduction of farming cultures. We can note that both authors refer to the figure of the "peasant" in their writings. While this is justified by the social and cultural context (e.g. in Switzerland most of the farmers still call themselves "*paysans/Bauer/contadino*", i.e. "peasant" in French/German/Italian), this category conveys a strong sense of belonging, tradition, and timelessness. Using it in academic publications, even reflexively, might not help in integrating a more dynamic and negotiated definition of identity and culture.

However, the conceptualisation of habitus and ethos as embodied social structures allows a better inclusion of the practical dimension of culture. For Droz and colleagues, as well as for Haggerty and colleagues, the farmers' ethos is expressed not only in narratives and discourses, but also and above all in farming practices, everyday actions, and physical symbols. This material and practical manifestation of the good farmer contributes to understanding how the cultural dimension of farming is anchored in farmers' everyday life and how it is more than a narrative on the profession and its morality. The 'good farmer' then is more about doing than talking or even thinking. This practical dimension might be weaker, even if not absent, in the social representation theorisation. Forney's conceptualisation is useful in highlighting the co-constructed (through interaction), negotiated (with the concept of posture), and dynamic (due to the differentiation between core and marginal elements) nature of the 'good farmer'. But its socio-psychology origins anchor this framing firstly in the things of the mind, even if the embodied and enacted dimension of representation is acknowledged.

The attention paid to practices is even clearer in Burton's theorisation of the symbolic value of farming practices, above all when he draws on Bourdieu's theory of capitals (Burton et al., 2008). By focusing on the competitive dimension of the 'good farmer', also emphasized by Haggerty et al. (2009), Burton highlights that production of 'good farmer' symbols is mostly achieved through actions that are open to scrutiny by the peer group and therefore unable to be manipulated. Consequently, this framing places the look and judgement of others at the centre of the identification to the 'good farmer', while Silvasti and Droz implicitly focus on farmer self-identification to the ideal. The idea of 'roadside' or 'hedgerow' farming illustrates clearly the importance given to social interaction and social control. Burton goes far deeper in presenting a detailed theorisation of the mechanisms at work in the identification to the 'good farmer'. The cultural dimension of farming becomes here more dynamic, interactive, and pragmatic. Moreover, it evolves with the appearance of new practices and new symbols and the vanishing of old ones. Interestingly, this theoretical development allows us to consider how governance policies could act with purpose on these social mechanisms, for instance in order to promote a cultural valorisation of environmentally friendly practices (Burton and Paragahawewa, 2011).

Although addressing the development of symbolic significance of farming practices in considerable detail and noting the transmission of cultural understandings from generation to generation, in Burton's work the processes of transmission (and transformation) remain hidden and relatively unexplored. In fact, he does not seek to outline the complete 'good farmer' as a person at all. Rather, he considers that much of farmers' social and culturally significant behaviour (the symbolically important production values) is mediated through an internalised 'good farmer' who, while often claimed to be external, is in fact a reflection of the self. This draws on his earlier use of identity theory with the symbolically important behaviours the equivalent of role behaviours that serve to define the identity of the farmer. It is by understanding how these behaviours come to symbolise the good farmer that, he contends, we will understand the cultural value of agricultural activities and, thus, the role culture plays in agricultural decision making. We believe that a more holistic view of the good farmer could be developed by integrating Burton's work on how good farming identities are constructed from farming activities with other concepts associated with the good farmer such as the moral and social role of the good farmer – along with developing a greater understanding of how the cultural values are transmitted between generations and between peers.

Finally, there is a fundamental difference in the way these diverse theoretical frames understand action in its relation to the cultural dimension. The Bourdieusian concept of habitus – while explicitly not ruling out conscious decision making (see Bourdieu, 1990) – gives predominance to embodied, subconscious behaviours that often escape the reflexive capacity of individuals. In contrast, conscious decision making and strategic behaviours are arguably more central to a symbolic interactionism approach. Here, individuals act more strategically, playing consciously with the rules defined by culture and social structures. While this opposition is a classic of social theory debates, the literature on the 'good farmer' does not address it centrally. One explanation could be that both the focus on the theory of capital (on the Bourdieusian side) and the attention paid to cultural scripts and identity (on the interactionist side) have contributed to mitigate the tension between structure and agency. Indeed, capitals are used by actors to play in a social field, and this implies some degree of strategy and agency, while a certain level of determinism underlies the concepts of scripts and identity. Forney's (2010) use of the social representation framing can be understood as an attempt to integrate social interactions in the making of cultural values. However, Emery (2010) probably addresses more directly this tension between structure and agency in his theoretical discussion around the rhetoric-culture approach, highlighting the importance of interactions and a creative use of values and culture in an "incessant play of agents and patients doing-and-being-done-to in culturally transient situations" (p. 25–26).

To understand theorisations underlying the good farmer construct it is important to recognise, as noted in the introduction, that the 'good farmer' is not a theory per se but an analytical construct – one that needs to be underlain by sociological or behavioural theory in order to be of value to researchers beyond its meaning as a 'common sense category'. The multitude of different approaches to theorising the 'good farmer' illustrates how the concept of the good farmer emerges from a multitude of literatures and in a multitude of ways. They all have in common a rejection of a normative stance (i.e. the author defines what are the characteristics of a good farmer), and all seek to develop an understanding of the cultural dimension of farming that goes beyond its descriptive synthesis, engaging with the significance and mechanisms of culture in farming.

Our analysis in this chapter revealed that there are two key components to good farmer theorisations. The first is that of cultural transmission – where the 'good farmer' is perceived to be the result of socialisation into agriculture and thus evolves from the transmission of established cultural beliefs and values. The famed, almost mythical, 'traditional' nature of agriculture would appear to support this notion. However, without explaining how cultural values and beliefs evolve over time, we contend a pure "cultural transmission" approach, such as cultural scripting, is of limited value to understanding the changing identity of the 'good farmer'. The second dimension sees the good farmer identity as developing largely through the practice of agriculture. Here the focus is more on the practice-based development of cultural significance and in particular, how the need to express aspects of the good farmer that are culturally significant (e.g. generate cultural and symbolic capital) affects the adoption of new roles and behaviours. This too can be problematic where issues of cultural transmission and non-production-related facets of the 'good farmer' identity, such as morality and social obligation, are ignored. To evolve as an analytical construct, we contend that more attention needs to be paid to theorisations that integrate these two dimensions.

As a final note, we would like to highlight the fact that, so far, theorisations of the 'good farmer' have remained strongly rooted in a 'representational' paradigm, as part of a wider 'cultural turn' in rural and agricultural studies. Since it was first developed however, novel approaches have emerged in social sciences that have resulted in what has been labelled as a 'non-representational' or 'more-than-representational' turn (see Lorimer, 2005; Vannini, 2015). As we will develop in Chapter 8, these approaches might offer stimulating opportunities to rethink the conceptualisation of the 'good farmer' that would pay more attention to the role of non-human actants, 'non-representational' (e.g. sensuous or affective) aspects in the construction the 'good farmer', or reframe it as a co-production resulting from multiple interactions within complex agricultural and food assemblages. However, this kind of theorisation of the good farmer has yet to be developed.

Notes

1 For a detailed analysis of Bourdieu's work, see Jenkins (1992) and for original works see Bourdieu (1986, 1990, 1998).

2 Authors' translation from the French: « une forme de connaissance socialement élaborée et partagée ayant une visée pratique et concourant à la construction d'une réalité commune à un ensemble social »

References

Abric, J.-C. (1994). L'organisation interne des représentations sociales: système central et système périphérique. In C. Guimelli (Ed.), *Structures et transformations des représentations sociales* (pp. 73–84). Neuchâtel and Paris: Delachaux et Niestlé.

Bourdieu, P. (1984). *Distinction: A social critique of the judgement of taste*. London and New York: Routledge.

Bourdieu, P. (1986). The forms of capital. In J. Richardson (Ed.), *Handbook of theory and research for the sociology of education* (pp. 241–258). New York: Greenwood.

Bourdieu, P. (1990). *The logic of practice*. Cambridge: Polity Press.

Bourdieu, P. (1998). *Practical reason: On the theory of action*. Stanford: Stanford University Press.

Bourdieu, P. (2008). *The bachelors' ball. The crisis of peasant society in Béarn*. Chicago: University of Chicago Press.

Burke, K. (1969). *A rhetoric of motives*. Berkeley, Los Angeles, London: California University Press.

Burton, R. J. F. (1998). *The role of farmer self-identity in agricultural decision making in the Marston Vale Community Forest*. Unpublished PhD Thesis. DeMontfort University, Leicester.

Burton, R. J. F. (2004a). Seeing through the 'good farmer's' eyes: Toward developing an understanding of the social symbolic value of 'productivist' behaviour. *Sociologia Ruralis, 44*(2), 195–215.

Burton, R. J. F. (2004b). Reconceptualising the 'behavioural approach' in agricultural studies: A socio-psychological perspective. *Journal of Rural Studies, 20*(3), 359–371.

Burton, R. J. F. (2012). Understanding farmers' aesthetic preference for tidy agricultural landscapes: A Bourdieusian perspective. *Landscape Research, 37*(1), 51–71.

Burton, R. J. F., Kuczera, C., and Schwarz, G. (2008). Exploring farmers' cultural resistance to voluntary agri-environmental schemes. *Sociologia Ruralis*, *48*(1), 16–37.

Burton, R. J. F., and Paragahawewa, U. H. (2011). Creating culturally sustainable agri-environmental schemes. *Journal of Rural Studies*, *27*(1), 95–104.

Burton, R. J. F., and Wilson, G. A. (2006). Injecting social psychology theory into conceptualisations of agricultural agency: Towards a post-productivist farmer self-identity? *Journal of Rural Studies*, *22*(1), 95–115.

Butler, D., and Holloway, L. (2015). Technology and restructuring the social field of dairy farming: Hybrid capitals, 'Stockmanship' and automatic milking systems. *Sociologia Ruralis*, *56*(4).

Carrithers, M. (2005a). Anthropology as a moral science of possibilities. *Current Anthropology*, *46*(3).

Carrithers, M. (2005b). Why anthropologists should study rhetoric. *Journal of the Royal Anthropological Institute*, *11*(3), 577–583.

Conway, S. F., McDonagh, J., Farrell, M., and Kinsella, A. (2016). Cease agricultural activity forever? Underestimating the importance of symbolic capital. *Journal of Rural Studies*, *44*, 164–176.

Coughenour, C. M. (1976). A theory of instrumental activity and farm enterprise commitment applied to woolgrowing in Australia. *Rural Sociology*, *41*(1), 76–98.

Dessein, J., and Nevens, F. (2007). 'I'm Sad to be Glad'. An analysis of farmers' pride in Flanders. *Sociologia Ruralis*, *47*(3), 271–292.

Droz, Y. (2001). Le paysan jurassien: un fonctionnaire qui s'ignore? *Journal des anthropologues*, *84*, 173–201.

Droz, Y. (2002). Du lait comme valeur. Ethnologie des fermes jurassiennes. *Ethnologie française*, *XXXII*(2002, 2), 209–219.

Droz, Y., and Forney, J. (2007). *Un métier sans avenir? La Grande Transformation de l'agriculture suisse romande*. Paris and Genève: Karthala/IUED.

Droz, Y., and Lavigne, J.-C. (2006). *Éthique et développement durable*. Genève and Paris: IUED/Karthala.

Droz, Y., and Miéville-Ott, V. (2001). *On achève bien les paysans. Reconstruire une identité paysanne dans un monde incertain*. Chêne-Bourg and Genève: Georg.

Droz, Y., Miéville-Ott, V., Forney, J., and Spichiger, R. (2009). *Anthropologie politique du paysage. Valeurs et postures paysagères des montagnes suisses*. Paris: Karthala.

Emery, S. B. (2010). *In better fettle: Improvement, work and rhetoric in the transition to environmental farming in the North York Moors*. Unpublished PhD. Durham University, UK, Durham.

Flament, C. (2007 [1989]). Structure et dynamique des représentations sociales. In D. Jodelet (Ed.), *Les représentations sociales* (pp. 224–239). Paris: Presses Universitaires de France.

Forney, J. (2007). Ethos et subversions quotidiennes chez les paysans romands. In S. Chappaz, A. Monsutti, and O. Schinz (Eds.), *Entre ordre et subversion* (pp. 119–134). Paris and Genève: Karthala/IUED.

Forney, J. (2010). *Produire du lait, créer du sens. Adaptations et résistances quotidiennes chez les producteurs de lait suisses romands*. (Thèse de doctorat Thèse de doctorat). Université de Neuchâtel, Neuchâtel. http://doc.rero.ch/record/18279?ln=fr

Forney, J. (2011). Idéologie agrarienne et identité professionnelle des agriculteurs: la complexité des images du "paysan suisse". *Yearbook of Socioeconomics in Agriculture, 2011*, 13–33.

Glover, J. L. (2010). Capital usage in adverse situations: Applying Bourdieu's theory of capital to family farm businesses. *Journal of Family and Economic Issues*, *31*(4), 485–497.

Gustavsson, M., Riley, M., Morrissey, K., and Plater, A. J. (2017). Exploring the socio-cultural contexts of fishers and fishing: Developing the concept of the 'good fisher'. *Journal of Rural Studies*, *50*, 104–116.

Haggerty, J., Campbell, H., and Morris, C. (2009). Keeping the stress off the sheep? Agricultural intensification, neoliberalism, and 'good' farming in New Zealand. *Geoforum*, *40*(5), 767–777.

Herman, A. (2015). Enchanting resilience: Relations of care and people – place connections in agriculture. *Journal of Rural Studies*, *42*, 102–111.

Howden, P., and Vanclay, F. (2000). Mythologization of farming styles in Australian broadacre cropping. *Rural Sociology*, *65*(2), 295–310.

Hunt, L. (2010). Interpreting orchardists' talk about their orchards: The good orchardists. *Agriculture and Human Values*, *27*(4), 415–426.

Huttunen, S., and Peltomaa, J. (2016). Agri-environmental policies and 'good farming' in cultivation practices at Finnish farms. *Journal of Rural Studies*, *44*, 217–226.

Jenkins, R. (1992). *Pierre Bourdieu*. Key Sociologists Series. Abingdon, UK and New York: Routledge.

Jodelet, D. (Ed.). (2007 [1989]). *Les représentations sociales*. Paris: Presses Universitaires de France.

Lorimer, H. (2005). Cultural geography: The busyness of being 'more-than-representational'. *Progress in Human Geography*, *29*(1), 83–94.

Mather, A. S., Hill, G., and Nijnik, M. (2006). Post-productivism and rural land use: cul de sac or challenge for theorization? *Journal of Rural Studies*, *22*(4), 441–455.

McGuire, J., Morton, L., and Cast, A. (2013). Reconstructing the good farmer identity: Shifts in farmer identities and farm management practices to improve water quality. *Agriculture and Human Values*, *30*(1), 57–69.

Mead, G. (1934). *Mind, self, and society: From the standpoint of a social behaviouralist*. London: University of Chicago Press.

Ploeg, J. D. van der. (1994). Styles of farming: An introductory note on concepts and methodology. In J. D. van der Ploeg and A. Long (Eds.), *Born from within: Practice and perspectives of endogenous rural development* (pp. 2017–2227). Assen, The Netherlands: Van Gorcum.

Ploeg, J. D. van der. (2003). *The virtual farmer. Past, present, and future of the Dutch peasantry*. Assen, The Netherlands: Royal van Gorcum.

Raedeke, A. H., Green, J. J., Hodge, S. S., and Valdivia, C. (2003). Farmers, the practice of farming and the future of agroforestry: An application of Bourdieu's concepts of field and habitus. *Rural Sociology*, *68*(1), 64–86.

Riley, M. (2016a). How does longer term participation in agri-environment schemes [re]shape farmers' environmental dispositions and identities? *Land Use Policy*, *52*, 62–75.

Riley, M. (2016b). Still being the 'Good Farmer': (Non-)retirement and the preservation of farming identities in older age. *Sociologia Ruralis*, *56*(1), 96–115.

Riley, M., Sangster, H., Smith, H., Chiverrell, R., and Boyle, J. (2018). Will farmers work together for conservation? The potential limits of farmers' cooperation in agri-environment measures. *Land Use Policy*, *70*, 635–646.

Rogers, M., Barr, N., O'Callaghan, Z., Brumby, S., and Warburton, J. (2013). Healthy ageing: Farming into the twilight. *Rural Society*, *22*(3), 251–262.

Saunders, F. P. (2016). Complex shades of green: Gradually changing notions of the 'Good Farmer' in a Swedish context. *Sociologia Ruralis*, *56*(3), 391–407.

Shortall, O., Sutherland, L.-A., Ruston, A., and Kaler, J. (2018). True cowmen and commercial farmers: Exploring vets' and dairy farmers' contrasting views of 'Good Farming' in relation to biosecurity. *Sociologia Ruralis*, *58*(3), 583–603.

Shucksmith, M. (1993). Farm household behaviour and the transition to post-productivism. *Journal of Agricultural Economics*, *44*(3), 466–478.

Silvasti, T. (2003a). The cultural model of "the good farmer" and the environmental question in Finland. *Agriculture and Human Values*, *20*(2), 143–150.

Silvasti, T. (2003b). Bending borders of gendered labour division on farms: The case of Finland. *Sociologia Ruralis*, *43*(2), 154–166.

Stock, P. V. (2007). 'Good Farmers' as reflexive producers: An examination of family organic farmers in the US Midwest. *Sociologia Ruralis*, *47*(2), 83–102.

Stryker, S. (1968). Identity salience and role performance: The relevance of symbolic interaction theory for family research. *Journal of Marriage and Family*, *30*(4), 558–564.

Stryker, S. (1980). *Symbolic interactionism: A social structural version*. London: Benjamin/Cummings.

Stryker, S. (1987). Identity theory: Developments and extensions. In K. Yardley and T. Honess (Eds.), *Self and identity: Psychological perspectives* (pp. 89–103). Chichester: John Wiley and Sons.

Stryker, S. (1994). Identity theory: Its development, research base, and prospects. In N. K. Denzin (Ed.), *Studies in symbolic interactionism 16* (pp. 7–20). London: JAI Press INC.

Sutherland, L.-A. (2013). Can organic farmers be 'good farmers'? Adding the 'taste of necessity' to the conventionalization debate. *Agriculture and Human Values*, *30*(3), 429–441.

Sutherland, L.-A. (2019). Agriculture and inequalities: Gentrification in a Scottish parish. *Journal of Rural Studies*, *68*, 240–250.

Sutherland, L.-A., and Burton, R. J. F. (2011). Good farmers, good neighbours? The role of cultural capital in social capital development in a Scottish farming community. *Sociologia Ruralis*, *51*(3), 238–255.

Sutherland, L.-A., and Darnhofer, I. (2012). Of organic farmers and 'good farmers': Changing habitus in rural England. *Journal of Rural Studies*, *28*(3), 232–240.

Swartz, D. (1997). *Culture and power: The sociology of Pierre Bourdieu*. Chicago and London: University of Chicago Press.

Thomas, E., Riley, M., and Spees, J. (2019). Good farming beyond farmland – Riparian environments and the concept of the 'good farmer'. *Journal of Rural Studies*, *67*, 111–119.

Tsouvalis, J., Seymour, S., and Watkins, C. (2000). Exploring knowledge-cultures: Precision farming, yield mapping, and the expert – farmer interface. *Environment and Planning A: Economy and Space*, *32*(5) 909–924.

Vanclay, F., and Enticott, G. (2011). The role and functioning of cultural scripts in farming and agriculture. *Sociologia Ruralis*, *51*(3), 256–271.

Vanclay, F., Howden, P., Mesiti, L., and Glyde, S. (2006). The social and intellectual construction of farming styles: Testing Dutch ideas in Australian agriculture. *Sociologia Ruralis*, *46*(1), 61–82.

Vannini, P. (2015). Non-representational research methodologies: An Introduction. In P. Vannini (Ed.), *Non-representational methodologies: Re-envisioning research* (pp. 1–18). New York: Routledge.

Verral, H. (1799). On the use and misuse of lime, with some observations and animadversions on other parts of farming. *Annals of agriculture and other useful arts* 32, 99–136.

Walford, N. (2003). Productivism is allegedly dead, long live productivism. Evidence of continued productivist attitudes and decision-making in South-East England. *Journal of Rural Studies, 19*(4), 491–502.

Weigert, A. J., Teitage, J. S., and Teitage, D. W. (1986). *Society and identity*. Cambridge: Cambridge University Press.

Wiederman, M. W. (2015). Sexual script theory: Past, present, and future. In *Handbook of the sociology of sexualities* (pp. 7–22). Switzerland: Springer.

Wilson, G. A., and Burton, R. J. F. (2015). 'Neo-productivist' agriculture: Spatio-temporal versus structuralist perspectives. *Journal of Rural Studies, 38*, 52–64.

5

MORALITY AND THE 'GOOD FARMER'

The many theorisations of the good farmer detailed in Chapter 4 illustrate how the concept of the good farmer has been used as an analytical tool by a variety of researchers and in a variety of ways. However, many of these theorisations – being targeted at specific research questions – do not attempt to present a complete picture of the good farmer him/herself. This differs from the concept of the good farmer as it appeared historically. Then notions of family, community, and morality were integrated within the good farmer to a far greater extent than is the case today – the good farmer was 'good' not only because of his/her ability to optimise agricultural production, but because he/she was morally an upstanding individual around whom communities were formed. In contemporary professionalised and specialised Western agriculture, farming has been framed mostly as a profession, implying a separation between professional skills and more general morality. However, two main dimensions of the 'good' remain intertwined in the 'good farmer'. First, 'good' is an evaluative term to describe a high level of competence in the performance of professional farming activities. This dimension has been at the core of the recent 'good farmer' literature which has focused mainly on activities of cultural significance. The second dimension is an evaluation of whether the behaviour of the farmer – in this case reaching beyond simple farming activities – is valuable for or harmful to society. This places the farmer in a wider social evaluative framework – it is not only his/her position in the peer group he/she is concerned about but also the role of farmers collectively in society. From this perspective, morality is a fundamental feature of farming.

In this chapter, we want to reintegrate this moral dimension into the scientific 'good farmer' conceptualisation by locating it in a discussion on the environmental impacts of farming. Indeed, nowadays many issues centre around the environmental 'morality' of agriculture, such as the conflict between the farmers' role as food providers for a growing population and the need to keep greenhouse gas emissions to a minimum, the role of agriculture in supporting rural communities, farmers' moral responsibility as stewards of the environment, farmers' moral responsibility for maintaining the welfare of their livestock, and/or, increasingly, the morality of providing economic support to farmers in the form of agricultural subsidies.

While moral issues permeate contemporary agriculture, we use this chapter to interrogate the existing good farmer literature for moments of tension when a hegemonic understanding of 'good farming' has been challenged by external threats (such as policy or economic change) that force a rethinking of the morality of agriculture. To that end, we first describe the issue of morality in agricultural practices. Second, we examine the good farmer literature to question what is 'good' about good farming – for the farmer, the farm, and the wider environment. Third, we engage in a dialogue between the good farmer literature and some of the recent literature on the social science of morality. In particular, we present Bell's (1994, 2018) natural conscience, Abend's (2014) moral background, and Justin Farrell's (2015) work describing moral orders as a means of providing us some theoretical language to address morality issues and conflicts within the 'good farmer' concept. We conclude by revisiting the good farmer literature with these new tools for understanding moral choice in agriculture.

The morality of agriculture

Attempts to claim the moral high ground in agriculture using the notion of the good farmer have been around for centuries. In the earliest English literature, as noted in Chapter 2, the good farmer was embedded within a religious context and, consequently, behaviours such as the fair treatment of servants, wives, family, and others in the community fell into the core moral framework of the 'good farmer'. The next moral claim to the good farmer came from the improvers in the 18th and 19th centuries. Wars around Europe (and corresponding food shortages) combined with rapid urbanisation made agriculture across Europe an increasingly nationalistic endeavour, if not a necessity. Production needed to be dramatically enhanced for wider society to survive. At the time, agriculture in Britain was dominated by small tenant farmers following customary practices and wealthy landlords focused only on rent generation – both of whom stood increasingly accused of immorality in failing to produce the food necessary to feed the country. This was exacerbated by rapid technological and scientific advances which made the gap between what could be produced and what was being produced ever wider. With a population regularly short of food, this growing disparity was easily portrayed as an issue of morality and the moral imperative of production was consequently driven into the 'good farmer' definition by those who controlled the debates.

A rapidly growing global population, continuous technological improvement, regular wars, and periodic famines have meant that the moral imperative of production is still part of the good farmer definition today as farmers continue to justify their practices as part of their duty to "feed the world" (e.g. Stuart and Houser, 2018). However, a shift occurred in the decades of the 1970s and 1980s, when it was recognised that the intensive agricultural practices necessary to optimise production were causing considerable environmental damage. The necessity of focusing only on production was further bought into question by an oversupply of agricultural produce in the 1980s – the development of butter mountains, milk lakes, and so on, meant that the justification of productivism on the basis of moral duty was lost. Instead, farmers found themselves in the position of receiving subsidies from the state to produce agricultural goods no one needed while, at the same time, causing increasing damage to the environment and lowering welfare standards. The agri-environmental and multifunctionality policies introduced to Europe at the end of the 1980s represented another attempt to define the morality of the good farmer – replacing the farmer producer with the farmer steward and community services provider (e.g. recreation, accommodation). Some fought against this, presenting arguments such as "farmers are not foresters" (Williams et al., 1994) and reinforcing the moral imperative of feeding the world as a justification for potentially damaging intensive agricultural practices (e.g. pesticide use – Duram, 1997; Stuart et al., 2012).

Agricultural and agri-environmental policies are often based on unexamined and unreflexive concepts of what is socially and politically meant by 'good farming' as they look to uphold certain farmers as good while excluding others as bad. Nonetheless, conflict over what is a 'good farmer' impacts how policies are written, what kind of farming is supported and engendered, what is seen as legitimate in farming communities, farmers' sense of self, and what is considered worth researching. Ultimately decisions about what a 'good farmer' is – where farmers are positioned on issues of morality – affect the success of such policies. Thus, the boundaries between who is considered a good (and, by extension, also bad) farmer become highly relevant not only for individual farmers and farms, but the wider field of farming as well as agricultural policies struggling to respond to a warming planet.

Today's agriculture is much more fragmented with farmers having numerous and different moral imperatives with which to anchor their farming and, consequently, many debates ensue about what the moral 'good farmer' should do. This chapter is about this conflict – of the times when symbols of what it means to be a good farmer are challenged both personally and collectively by people who hold different notions/symbols of the moral imperatives in agriculture. This could mean organic farmers, those who do not adhere to conventional methods, or new entrants who value outcomes that are different to those traditionally valued in a region, among many other possible examples.

What fosters changes in the definition of the 'good' in farming often revolves around economic changes in the region and the industry as well as the changing needs of wider society at the national level. While some authors have hinted at a moral basis of some of these changes and contradictions, academics have yet to engage fully with the social science of morality to explore what the 'good' in 'good farming' means.[1] As noted previously, for a few decades most of the moral discussion around farming, notably in European countries, has revolved around environmental issues. Riley (2016a, p. 63), following Bourdieu, makes clear the idea that the 'good farmer' is a useful way to understand the relationship between "farmers and conservation" and the resistance met by external pressures to alter what should symbolise good farming practices (such as agri-environmental schemes – Burton et al., 2008; Burton and Paragahawewa, 2011; Forney et al., 2018; Stock et al., 2014), as they undermine individual and collective autonomy to establish or maintain 'good farmer' symbols. In other words, while other moralities exist, the moral dimension of the 'good farmer' cannot be discussed in the current context out of its relationship with environmental considerations.

Constructing the 'good' in the good farmer

As with the other elements of the concept, the moral dimension in the good farmer is essentially about social and symbolic interactions. Narratives and practices expressing the 'good' are boundary making processes that develop along the separation between what is me or us, a self, and what is not, the Other. However, to be a good something refers also to a shared definition of what it is to be this something – it is not possible to understand what is 'good' in farming without referring to what is farming itself. To integrate this idea of a functional good, we briefly present the idea of the functional farmer as inspired by the work of MacIntyre (1984).

A conflictual 'good': the self and the other

As noted previously, the recent good farmer literature – with the exception of a few papers (including Riley, 2016b; Stock, 2007; Sutherland, 2013) – has largely avoided addressing the centrality of moral issues within the concept of the 'good farmer'. The existing conflicts are well recognised, but usually focus on the tension between farmers and non-farmers, represented in policies, as exemplified by the following quote from Burton (2004):

> Within modern agricultural landscapes there lie meanings developed through the performance of everyday farm tasks by members of the farming community that to the farmers speak of their toil and personal victory over the land yet to us may represent the excesses of the agricultural industry – of pollution, industrialization and the degradation of the countryside aesthetic. *(p. 197)*

This approach is consistent with the objective of demonstrating the existence of a shared repertoire of symbols and values among farmers. Arguably, however, it supports a binary conceptualisation of moral conflict by emphasizing the important role of non-farmers as the main 'Other'. This binary of Self and Other has its origins in the symbolic interactionist school of sociology typically associated with George Herbert Mead and Erving Goffman. In brief, to understand oneself as a unique person with an identity, a Self, one has to be aware of Others. I am myself because I am not that Other over there.

The consequences for our understanding of the 'good farmer' are directly related to the symbols and practices that define who is and who is not a 'good farmer', e.g. 'I am myself a good farmer because others recognise my choices, represented through the symbols of my farm practices, as emblematic of being a good farmer'. These are shared symbols and practices. We examine an extension of this theoretical line of thought later in Michael Bell's work and then look at how conflicts arise when the Self and the Other are at odds or in disagreement. For our purposes here, what is most striking is that because farmers share a rural sensibility of what is good – which is somewhere at the core of the good farmer concept – there is a tendency to look at moral conflicts essentially as falling between 'farmers' (as a uniform group) and non-farmers (urban bureaucrats, environmentalists, etc.). This has the effect of simplifying the notion of morality in agriculture, negating the moral conflicts that exist within the farming communities themselves and, in particular, failing to address the role of morality in the multiple, diverse, and shifting notions of the 'good farmer' (see Chapter 7). Within the 'good farmer' literature there is an opportunity to open up this binary concept to a more nuanced analysis of the various fault lines exhibited through certain practices and symbols indicative of certain moral frameworks.

Following Lowe et al. (1997) and Bryant and Garnham (2018), we view these competing moral value systems as both reinforcing the position of the Self (as a 'good farmer') and enabling the identification of the Other (as a 'good' or 'bad farmer', or 'non-farmer') – beyond the performance and competence dimension.[2] The concept of the 'good farmer' revolves around shared assumptions of what is good about a farmer or the outcomes of specific farm practices articulated through observable symbols. Often, in the use of the 'good farmer' term by farmers and social scientists (see Chapter 1) there is a presumption of a dominant style of farming that is deemed to be good. Yet, trends in farming belie this presumption. Within the last 30 years there have been significant debates on whether farming is becoming more multifunctional or neoliberal, more organic or more conventional, or less or more cooperative. Just these simple dichotomies indicate that what is 'good' is often both contested and unstable. Introducing diverging farming moralities is particularly important because policy solutions are often based on the assumption that major issues such as alleviating global hunger or addressing climate change can prima facie be addressed through technological solutions (Stock, 2015) based on the assumption that farmers constitute a single group of like-minded economically rational decision makers who respond to policy in the same way. Yet Gasson's (1973) work identifying the different goal structures of farmers, Ploeg's (1994) 'farming styles', Morris and Evans' (2004) description of 'agri-cultures', Burton and Wilson's (2006) finding of multiple and layered identity structures, and many other studies that have looked for structure in farming communities, all indicate that farming communities are infinitely more complex.

The articulation of a position of the Self and the identification of the Other in moral conflicts also allows us to integrate a dialogical and dynamic perspective in the understanding of farming moralities, where the role of the Other alternates between other farmers who are engaged in different practices and non-farmers expressing moral judgement over farming. Saunders (2016) and Sutherland (2013) both refer to identities being fragmented or under stress based on the changing contexts. As the field of farming changes over time – changes which, as we observe in Chapter 3, can be considerable – the notion of what it means to be a good farmer also changes as well, reflecting what Saunders (2016, p. 395) calls "disjointed norms". For example, Sutherland (2013) observes the existence of pragmatic organic farmers – conventional farmers who witnessed the financial stability of organic farmers and were persuaded to change some practices to follow suit. Similarly, Saunders (2016) describes how farmers will make changes even at the cost of some practice-based identity in order to remain economically viable. In another example, Coughenour (1976) observed that social processes involved in specialised commodity production played an important role in the development of social and moral norms for Australian sheep farmers. When an economic crisis hit the wool sector, some farmers converted to beef cattle production nevertheless – with the likelihood of conversion determined in part by the extent of identity commitment to sheep.

Within different forms of farming there are distinct sets of rules (norms/repertoires) and farming practices that, when applied, generate symbols of 'good farming'. However, when conflicting moral beliefs are attached to environmental management behaviours, the understanding of these practices and symbols may lead to inconsistency in response. What was once deemed environmentally good may become environmentally bad and thus morally bad. Belonging to a group helps shape these moral beliefs. When those beliefs are challenged a tension emerges between a desire to hold on to one's beliefs or to one's group. This can then result in a break of some sort – either in the person's sense of self (e.g. a self-identity as a sheep farmer is difficult to maintain when raising cattle) or in the person's group membership (e.g. a conventional farmer who converts to organic would have difficulty socialising with conventional farmers as a result of holding a different set of values and beliefs consistent to his/her new identity).

Moral philosophers sometimes refer to the good farmer to help illustrate a point they are making (Foot, 1969; Murdoch, 2001). After a discussion of the relationship between a watch and a good watch, MacIntyre (1984) extends the thinking to a farmer who gets a better yield for his crop per acre than any of his neighbours, has a very effective program of soil renewal, and wins all the first prizes at the agricultural show with his cattle. MacIntyre claims that it is valid to say that this man is a 'good farmer', because he does well what a farmer is supposed to do. If a watch can be defined as an object that gives the time, a good watch will give it in a better (more precise, for longer, etc.) way than others. For MacIntyre, the concept of the farmer is a functional concept as it defines a farmer, as with a watch, in terms of the purpose or function we attribute to a farmer (MacIntyre, 1984, p. 58).

A functional approach to the 'good' is therefore useful to overcome some of the limitations of a purely interactionist perspective that could reduce the 'good' to a process of boundary making between us and them. By referring to a fundamental definition of what it is to be a farmer it reminds us of the ontological dimension of the 'good farmer' concept.[3] In a discussion on the moral connection between the 'good farmer' and the environment, this functional definition of the 'good' contributes to expanding the discussion beyond stating diverging moral positioning around nature. It emphasizes the fundamental relationship between the definition of the 'good farmer' and an ontology of farming. To put it simply, following MacIntyre, disagreements about the practices and symbols that make a 'good farmer' always imply a wider discussion about what are the functions and purposes (as well as outcomes) of farming itself.

The good in recent social science

The 'good farmer' concept was initially based on farmer-defined notions of what it means to be a good farmer with the symbols of 'goodness' indicating factors such as competence, proper decision making, increasing yields, and profitability. The literature has addressed the clashes between these notions and others – for example, those implied by the development of agri-environmental schemes (and related programs) that utilise measures of ecosystem health to indicate 'good farming' practices. As Burton et al. (2008) observe, healthy natural ecosystems are inherently complex or even chaotic, whereas farmed landscapes are generally dominated by order and regularity. This discrepancy is a common reason for farmers resisting enrolling in such voluntary programs. These developments also both hint at externally driven notions of what a good farmer is and does and suggest tension between competing definitions of the 'good' – which emerge both as material outcomes and moral entanglements. To disentangle these ideas we turn to key social science of morality developments that have emerged in parallel with the good farmer literature. Three key theorisations are central to developing our understanding of 'good' in 'good farmer': Michael Bell's natural conscience (1994, 2018), Gabriel Abend's (2014) moral background, and Justin Farrell's (2015) moral orders.

The natural conscience, the natural self, and the three natures

US rural sociologist Michael Bell (2004, 2018) offers a significant body of conceptual work on the morality of ruralness with a focus on agriculture. For our purposes, it is important to put our wider discussion of the good farmer concept in dialogue with Michael Bell's work as the two have developed in parallel without much direct connection. Part of the reason for this stems from the more Europe-centric focus of the good farmer literature as opposed to Michael Bell's more America-centric work. Most importantly, however, Bell provides some theoretical concepts that enable us to develop a wider theoretical lens around the 'good' in the 'good farmer'. Drawing primarily on a sociological theoretical tradition, Bell brings Emile Durkheim's collective conscience and Mead's symbolic interactionism to bear on our contemporary understandings of the relationship between morality and the environment related to farm practices.

The 'natural conscience' is what Bell (2018, p. 72) describes as "a basis for moral thinking we believe to be free of society and all its politics and constant play of interests and ambitions". Bell uses the phrase as a companion to Durkheim's collective conscience. The natural conscience is based on the belief of the existence of a pure and authentic nature that would pre-exist the social – what Bell refers to as "first nature". Bell related this natural conscience to the interactionist theory of the Self. The natural conscience emerges from the co-constitutive notions of the natural me and the natural Other, both thought of as pure, real, and true. As Bell (2018) puts it:

> Thus we feel the natural me as a more authentic and valid self than what society creates, sees, and guides. The natural me [Self], then, we imagine as a non-political self, as opposed to the self that emerges from society's constant tussle and scuffle of interests and ambitions.
>
> *(p. 73)*

The reference to a notion of nature as pure, the unspoiled and uncorrupted by the social and its politics, creates a sense of a fundamental goodness of the natural Other and the natural Me. The idea of the natural being not only first, but fundamentally 'good' refers to what Bell names "second nature" or "nature as moral good".[4] For this precise reason, as Bell continues, we see the natural conscience as being a "more sure basis for the moral" (p. 73). From the natural Me and natural Other also derive the collective self, "a sense of non-political community derived from a natural other" (p. 138), and notions of a natural We and a natural Them, that are essential in the politics of morality. The natural We builds on the idea that some others are "similarly constituted to me" (p. 226). The hidden politics of the 'natural' appear again in the tensions and conflicts that might appear between a natural We and a natural Them. The supposed goodness associated to the natural Other and natural Me expands to the natural We in opposition to a natural Them, as a way to legitimate and protect common advantages. This is what Bell terms as the "use of nature in nonpolitical politics" (p. 226), highlighting the generally bounded and hierarchical dimension of the natural We that serves to defend unbalanced power relations, domination, or other advantages, by placing them in the different nature of the natural Them.

Bell's works on the moral basis of first nature provides useful elements to think through the moral conflicts in farming. First, the concepts of natural conscience and natural Me highlight how moral values – for instance in relation to farming and the environment – are deeply rooted in the way people (e.g. farmers) understand goodness in relation to who 'they truly are'. This offers an interesting echo to the ontological dimension of the 'good farmer', described previously as inspired by

MacIntyre's notion of a functional concept. Second, the intimate conviction of a natural Me/We being fundamentally good resonates with Silvasti's (2003) observation of Finnish farmers genuinely believing that acting according a 'good farmer' script could not be harmful to the environment. Finally, the non-political politics contributes to the understanding of how a sense of moral legitimacy and justice can be developed by promoters of diverging definitions of the good farmer, by drawing on 'natural' and innate qualities, positive for them or negative for their opponents.

This phenomenology of farming (where the practice of farming is both shaped by moral ideas and shapes what is understood to be 'good' at the same time) that integrates tensions and a never-ending struggle for meaning (Bell, 2004, pp. 121–122) allows us to address morality and identity in relation to the good farmer. However, Bell (2018), in his wider-lens work on the 'good' in relation to our understandings of the environment, offers a significant caution for those seeking to understand the good in people and things, namely:

> The moral philosopher in all of us wants to find a sound basis for the good. On the whole, this is a welcome desire. But the sociologist in all of us should also recognize that we need to tread carefully with eyes wide open, and with mouth and ears ready to give and receive critique. The good, when we treat it as an absolute, is so seductive we may hardly notice when we are using it to do quite bad things – what I call the conundrum of the absolute.
>
> *(p. 224)*

In other words, Bell invites us to welcome and combine both a moral and absolutist approach to the good (like moral philosophers) as a necessary guidance for action – *and* a more relativist and critical stance (like social scientists) that ponders power relations and manipulations of the good in the politics of, in our case, farming. Recent work on the 'good farmer' develops some of this critical realisation of changing contexts/boundaries (Riley, 2016b; Saunders, 2016; Sutherland, 2013). While Bell's natural conscience, natural Me/We, natural Others/Them, and interpretations of 'nature' help us clarify some of what might be considered good farming, Gabriel Abend's work on moral backgrounds helps to clarify some of the moral foundations that make such judgements socially and culturally significant.

First and second order morality and the moral background

Abend (2014), drawing on examples from US business ethics, makes a distinction between first and second order morality and the moral background. First order morality comprises the behaviours and actions taken on behalf of those making ethical choices. These actions are then judged to be good or bad by others – and it is the act of judging the morality of others that comprises second order morality. This can be illustrated through an example from agriculture. In New Zealand dairy farming one farmer might choose to put up fencing to keep cows out of the waterways while another might choose not to put fencing up (Haggerty et al., 2009; Tall and Campbell, 2018). The choice to put fencing up or not, for Abend, is part of the first order morality sometimes called ethics. Farmers' judgements about each other's choices and the judgements of all others comprise second order morality. We can conceptualise morality in relation to the good farmer as a situation where effects on the environment are involved in those judgements – as in the case of Bell's third interpretation of nature as a moral 'bad'. Abend (2014, p. 53) furthers our understanding by describing the moral background that "facilitates, supports, or enables morality". Specifically, the moral background is "a particular collection of para-moral elements" that include what is or is not able to be morally judged, the cultural repertoires available, and how moral judgments are shared and debated along with metaphysical assumptions (Abend, 2014, p. 67). Obviously, moral background as well as first and second order morality vary according to cultural and geographic contexts.

Abend (2014, p. 37) argues that social actors have access to a repertoire of moral concepts which "enables and constrains their thought and speech, their laws and institutions, and importantly, the actions they may undertake". In farming, these repertoires typically play out in embodied enactments of what is culturally agreed upon as 'good' given the farming culture a farmer finds his/herself within; moreover, they point to what is visible and included and what remains invisible, overlooked, or ignored in this process (Abend, 2014, p. 55, fn 53; Carolan, 2011). For us, Abend's key observation is that moral backgrounds shape farm practices in pursuit of what is 'good' in farming by helping establish the boundaries of judgement based on cultural repertoires.

Competing moral orders and moral devaluation

Farrell's (2015) work offers a framework for analysing competing notions of what is 'good' with a specific emphasis on place and the environment. Farrell, in his exploration of the Greater Yellowstone ecosystem in the western United States, developed the concept of competing moral orders based on his analysis of competing understandings of what it means to live a moral life in the US West. In doing this he introduces an important concept that partners well with Abend's and Bell's work, namely the questions of conflict, moral disagreement, and the politics of morality. For Farrell (2015, p. 10), drawing on a sociology of culture (especially Wuthnow, 1989), moral orders "are an interpersonally and institutionally shared structure of moral beliefs, desires, feelings and boundaries that are derived from larger narratives and rituals". In the same way as judging one farmer's actions as bad and another's good (second order morality for Abend), moral orders help shape and reinforce who the Other is, while reinforcing self-belief about what is 'good' in one's own actions. With an emphasis on the moral being indicative of something right, good, worthy and just, Farrell argues that these moral orders orient actions towards each other in workplace, family, public, and political engagements. He further contends that they help give "narrative coherence around an individual's life" (p. 11). The coherence aspect shows similarities with the good farmer literature as it ties one's success as a good farmer to the strength and dynamism of one's identity as a farmer. In the context of the Greater Yellowstone ecosystem, Farrell argues that the environmental conflicts at the heart of the major disagreements come down to conflicts over moral orders. Thus, as participants in favour of re-wolving the West gather evidence to support their beliefs, they also aim to devalue not only the beliefs of those who oppose wolf reintroduction, but their opponents' selves. As moral orders orient our personal narratives and self-worth, these conflicts at the level of belief undermine the opponents' self-worth – with the process of moral devaluation making compromise nearly impossible.

The concept of moral order is useful to express variations in the definition of good farming – which depends on environmental factors such as place, water availability, climate, topography, but also social aspects like technology, tradition, networks, infrastructure, or diet, among other things – in moral terms. Farrell's definition of moral order also helps to highlight that conflict over farming practices might be related to moral framings (notably for non-farmers) that are broader and largely unconcerned with how agriculture is actually practised. Moral orders help orient people's lives towards teleological goals. There is a specific point to one's life; life has a purpose and, with the concept of moral orders in mind, helps to relate the good farmer to this fundamental (but variable) moral dimension of farming – to feed the world, to keep the land in 'good heart', to pass the farm on to the next generation. Finally, the process of moral devaluation again emphasises that, in conflicts over the morality of farming practices and over the definition of the good farmer, what is at stake is not only the ambition of being an excellent farmer, it is mainly about self-worth and moral legitimacy.

Yet, disagreements over what is considered a good farmer occur (Sutherland and Darnhofer, 2012; Kessler et al., 2016). So how do the concepts of natural conscience, moral background, and moral order help us to understand contradictions in one's farmer identity (e.g. "I believe I'm a good farmer, but why are the waterways so polluted?") and between farmers with different understandings of what is good?

Incorporating social science of morality into the good farmer

As observed in the previous chapters, the existence of good farmers and specific farming cultures can be read from the symbols and landscapes they produce – symbols that, while they may provide evidence of moral behaviours to those within the peer group, may be read as something entirely different by those outside. Lowe et al. (1997) suggest that when farming cultures come under attack – whether from outside or from a competing sub-cultural definition of what a *good* farmer is – then a process of moralisation takes place that reinforces the existing culture (drawing the group together to resist the Other) while also altering it (making farmers question and re-establish the morality of their actions). Lowe and colleagues' example describe how dairy farming in Devon (England) went through a process of industrialisation in the 1990s. To that end farmers (and rural neighbours) grew to accept that some amount of water pollution (from fertiliser, manure, etc.) was necessary. By adapting the symbols of good/bad farming in this manner, even farmers that polluted to a certain extent were able to think of themselves and be thought of by others (using their second order morality lenses) as a good farmer.

Being considered a good farmer reconciles what it means to be a good person with what it means being a (good) steward. The conflicts that Lowe et al. (1997) observe can be described as competing groups with differing moral orders – the farmers (as a culture) and the encroaching outsiders (more urban, non-farming environmental groups, and changing government policies) – which offer different perspectives on the relationship between productivism and stewardship. For farmers, who were previously thought of as 'good' no matter what (due to a natural virtuosity inherent in the profession of farming), the reality was that making a living as a farmer necessitated high levels of production – in Bell's words, a concession to the economy that was not enough to contest the goodness of a 'natural We'. Outside agriculture, growing environmentalism accompanied by decreasing belief in agricultural exceptionalism has led many to view farmers as naturally 'bad' rather than naturally 'good', based on the extent of environmental damage caused by polluting agricultural practices. This parallels Farrell's (2015) idea of moral devaluation, i.e.

> [increasing public sentiment towards the environment] was leading to growing intolerance of the problems farmers faced. On this issue, therefore, farmers' personal sense of worth (in the stewardship ethos) faced social rejection, and inevitably this reinforced their feeling of isolation and alienation.
>
> *(p. 202)*

While the Devon dairy farmers were able to retain a sense that what they were doing was indicative of being a good person, at that time, farmers in England felt under siege from the changing moral perception of agriculture. As Lowe et al. (1999) observe:

> For farmers, *the morality of farm pollution concerns the morality of the deed* – whether the pollution was deliberate or accidental. Their sense of personal worth, responsibility and circumstances all come into play around the morality of farming and the morality of pollution.
>
> *(p. 204, emphasis original)*

An example of this moral devaluation can also be seen in Harrison et al.'s (1998) work on the Wildlife Enhancement Scheme in East Sussex which looked at farmers' and local residents' notions of conservation in agriculture. Here the observation that farmers were being paid to manage land in a manner that appeared destructive led many local residents to "question their original understanding of most farmers on the Levels as benign custodians of nature" (p. 316). They also drew a distinction between 'good' farmers who respected their land and wildlife, and 'bad' farmers – progressively seen as a 'natural Them' – who were driven by greed and disregarded nature and, as a result, considered that only some farmers could be trusted to deliver conservation. Farmers, on the other hand, invariably constructed themselves as active stewards of nature.

These examples illustrate how changing moral boundaries of good farming in a given region can create tension between farmers and the public, different kinds of farmers, farmers and the regional political structure, and farmers and the extant policy. Many authors, for example, Kessler et al. (2016) and Lowe et al. (1997), highlight the inherent contradiction of farmers seeming to simultaneously seem care for and pollute the land. Saunders (2016) argues that this tension might well indicate change happening in the field of agriculture:

> While it seems clear that the habitus of some farmers is still firmly aligned with productivist notions, which may retard prospects for rapid change, this does not rule out prospects for a more gradual change as these farmers circumspectly position themselves in relation to a multifunctional agriculture.
>
> *(pp. 402–403)*

Given that the previous considerations emphasize how deeply morality can be rooted in identity and self-worth, is it possible to become a different kind of good farmer? Riley (2016a, p. 65) quotes Bourdieu to emphasize that changing habitus in farming might need a "form of radical conversion" (Bourdieu, 1993, p. 78). Riley (2016a) incorporates a parallel literature on knowledge cultures which contends that negotiations between farming knowledge cultures and those of outsiders are infused with moral judgements (see Tsouvalis et al., 2000) and suggests that policy makers need to be aware of the moral sphere of those who consider themselves good farmers. Looking at different knowledge cultures in the development of agri-environmental schemes in the United Kingdom, Morris (2006, p. 125) reveals that "the porosity of the boundary between state-led and farmer approaches to knowing nature" facilitates knowledge negotiation. On this basis, she suggests:

> The opportunities that exist for productive interchange and negotiation between the knowledge cultures brought to bear on the environment by farmers and AES is a significant finding. It suggests that farmers views and understandings are being respected and accommodated and in this situation more effective environmental management is likely to ensue.
>
> *(p. 126)*

Bell's (2004) work on the organisation Practical Farmers of Iowa describes the relationships between farmers that engendered a shift in farming practices. This change – termed an epiphany by Bell – often follows a crisis of some kind, whether in farming when the rules change (Riley, 2016a, p. 65; Johnsen, 2004) or in their self and their sense of what is good. Linking to the broader concept of the good farmer, this thinking pays attention to the 'rules of the game' – but is useful in acknowledging that "rules, norms, and values are themselves not given but prone to contestation and reconstruction by members of the group or when challenged by 'outsiders'" (Tsouvalis et al., 2000, p. 912). The key issue here is that in order to meet others' moral expectations of what is 'good' in farming, farmers do not readily change their identity but, rather, find new "systems" to make what they were already doing coincide with their own identities of goodness (Riley, 2016a, p. 71).

Bell and Farrell illustrate how competition plays out. There is significant difficulty in moving from one group of good farmers to another – or what Bell would describe as becoming a different Self and joining a different group of Others. Initially, the shift has considerable social costs as "failure to display symbols of group belonging can result in social disapproval, leading to a sense of loss for the individual and a corresponding decrease in self-esteem" (Burton, 2004, pp. 198–199). This sense of loss and disorientation was also observed in farmers exiting from a productivist paradigm in Iowa in Michael Bell's ethnography of the Practical Farmers of Iowa – abandoning productivism and adopting sustainable practices was accompanied by a palpable feeling of social isolation (Bell, 2004, p. 118). Similarly, Bell observed that farmers who lost their farm during what is referred to as the '1980s Farm Crisis' lamented both the loss of status in the community (and of the farm itself) and, for many male farmers, a blow to their masculinity. Bryant and Garnham (2018) similarly observe that farmers accepting packages to exit from farming in Australia experience it as a moral event, laden with judgements, recriminations, and blame that revolve around these repertoires, natural consciences, and second order morality as "to accept a package transgresses codes of conduct in the ordinary ethics of farming communities." (p. 63) They suggest that moralising by the wider farming community "creates a moral distance between the self and Other" (p. 66) which reinforces their ability to view themselves as 'good farmers'.

This can be seen by examining contemporary literature from a moral perspective. In one of the earliest papers, Silvasti (2003) explored Finnish peasant farming using the concept of a 'cultural script' – an embedded cultural construct similar to Bourdieu's habitus (see Chapter 4). In Silvasti's (2003, p. 148) data there is an emergence of challenges to productivism within the peasant cultural script, but not enough for organic farmers to "put themselves apart from other farmers". In this case, organic farmers were reluctant to accuse non-organic farmers of polluting the environment because to do so would be "fouling their own nest" – in other words, while they recognised the position on the environment was different to their own, they still saw themselves as falling within the overall farming community, a same 'natural We'. The difference between peasant farmers' interpretations of the environmental risks of farming and those of the scientists was, Silvasti asserts, one of the most fascinating themes of the research.

The strength of the farming culture is also illustrated by Sutherland and Darnhofer's (2012) paper on organic farmers and the process of change related to the concept of the good farmer. While interviewing 'pragmatic' organic farmers (those that converted for primarily economic rather than ideological reasons), they found that organic farmers helped shift the dialogue around who was a good farmer and what that meant without viewing the organic farmers as having a different farming habitus – they were in the same 'field' in Bourdieusian terms. There exists a constant re/negotiation between the Self and the Other as farmers struggle to maintain a sense of identity among shifting policies, fads, and technological changes. While often still dismissed by conventional farmers as 'not real farming' or 'pretend farming' (Silvasti, 2003), the organic farmers in Sutherland and Darnhofer (2012) respected the conventions of farming enough to not "foul their own nest". Yet, the dramatic changes and upheavals to farming in the late 20th and early 21st centuries towards the environment as a core moral concern had clearly changed what had been stable "rules of the game".

The disruption to the feel for the game also opens up the potential for farmers to adopt different interpretations of what it means to be a good farmer (Haggerty et al., 2009, p. 769) – innovating creatively in response to major disruptions (such as bovine spongiform encephalopathy or 'mad cow disease') while still considering themselves to be conventional farmers (Sutherland and Darnhofer, 2012, pp. 235–236). Whereas Sutherland and Darnhofer (2012) and Silvasti (2003) witnessed an attempt to integrate change through the development of competing definitions of what is to be a good farmer thanks to a shared reference to a singular farming culture under duress, others articulate more fundamental breaks and interpret them as fragmentation into multiple farming cultures with an emphasis on plurality and the opposition between multiple ideas of what a good farmer might be (Bell, 2004; Johnsen, 2004; Haggerty et al., 2009).

Haggerty et al. (2009) explicitly discuss how changing contexts (particularly policy regimes) influences farmers' subjectivities and thus their understanding of what it means to be a good farmer. Here we can see Bell's ideas coming into play. Where Silvasti describes that "The peasant-oriented farming style is self-evidently thought to be ecological" (2003, p. 149), we see the natural conscience (Bell, 2018) at work. That farmers are reluctant to "foul their own nest" confirms that the "Denial of agro-environmental problems is consistent . . . with the cultural script of good farmerhood" as Silvasti (2003, p. 149) concludes. Farmers are reluctant to adopt a critical stance – a second order morality – against largely consensual farming practices as it could undermine what is culturally representative of a good farmer, and by this undermine the moral goodness of farming in general. Rather than organic farmers pressing a dialogue on what is a good farmer (even though most conventional farmers see organic as pretend or not real) (Silvasti, 2003, p. 146) to highlight these contradictions, the Others that challenge this denial are those coming from a different environmental morality, natural Others.

These observations represent only a fraction of the moral boundaries that serve to define what it means to be a good farmer. While here we focus on environmental boundaries, as we noted, other moralities in agriculture such as the conflicting moral imperative to 'feed the world' offer other avenues for exploration. Moral boundaries that define different sub-cultural notions of the 'good farmer' and the moral boundaries that divide farming and non-farming populations offer other possibilities. Some work on this already exists. Researchers have described differences between different farming cultures and sub-cultures (Bell, 2004; Naylor et al., 2018; Shortall et al., 2018; Stock, 2007) while others describe instances where farmers fall afoul or their rural neighbours (Lowe et al., 1997; Naylor et al., 2018). Carolan (2020) takes a further step towards incorporating morality and agriculture by theoretically connecting ethical or good consumers and the good farmer literature via a moral economy framework. By incorporating the social science of morality as suggested here we can better theorise the boundary zones between different good farmer identities and thus explore changing common sense notions of what it means to be a good farmer. An improved understanding will help us to both understand farmers' reactions to externally imposed policy measures (such as agri-environmental schemes) and assist in the development of effective policies in the future.

The importance of morality or what is good in farm practices has received less attention than the actual practices that represent a cultural consensus on what is a 'good farming'. Yet, as we have seen in this chapter, from the earliest times the issue of morality (or multiple moralities) has played an important role in determining who is a 'good' farmer. The addition of theoretical tools for interrogating what is meant by 'good' or moral in the phrase 'good farmer' provides a useful extension to the good farmer concept – offering us language and concepts that enable us to better identify the moral aspects of changing farm practices that are affecting farmers and agriculture worldwide. The chapter thus better positions the good farmer concept and scholars to be able to address the tensions of injustice, inequity, and conflict related to farm practices choices and identity formation. This issue is important for the next chapter, where we seek to understand the role of marginalised 'Others' in the 'good farmer' concept – focusing, in particular, on gender and sexuality. How do alternative moralities contest inequalities that are reproduced by dominant cultural definitions of agriculture? What happens when marginalised groups challenge the moral boundaries and moral assumptions of who is a 'good farmer'?

Notes

1 It is important to emphasize that studying moral differences between people – as the social science of morality does – does not require taking a normative stance. A researcher, by studying moral conflicts, does not necessarily support a specific moral norm as correct or right.
2 Naylor et al.'s (2018) contribution to the good farmer literature offers an update to the theory of planned behaviour with an emphasis on social identity theory and self-categorisation theory that attempts to measure the potency of one's identity as a farmer in given contexts.
3 For a wider discussion of the importance of ontology to what it means to be good farmer see Campbell (2020).
4 Bell also develops a third conception of nature, "nature as a moral bad", or nature as the seat of desires (and therefore politics) that should be controlled and tamed (p. 141). The opposition between second nature – the idea of a good nature – and third nature – opens up two developments that we will just mention here. The first relates to the internal dilemma about what should guide our actions, or in Bell's words: "Should we strive only to commit 'natural' acts or should we resist demon nature?" (p. 142). The second replaces

these two concepts of nature in two opposite moral positions towards natural processes, ecosystems, and the environment in general: one sees them as an unquestionable good; and another considers human action to control, tame, or improve natural forces as something inevitable, needed, or even as a moral duty.

References

Abend, G. (2014). *The moral background: An inquiry into the history of business ethics*. Princeton, NJ: Princeton: Princeton University Press.

Bell, M. M. (1994). *Childerly: Nature and morality in a country village*. Chicago, IL: University of Chicago Press.

Bell, M. M. (2004). *Farming for us all: Practical agriculture & the cultivation of sustainability*. University Park: Penn State Press.

Bell, M. M. (2018). *City of the good: Nature, religion, and the ancient search for what is right*. Princeton, NJ: Princeton University Press.

Bourdieu, P. (1993). *Sociology in question*. London, Thousand Oaks and New Delhi: Sage.

Bryant, L., and Garnham, B. (2018). Farming exit and ascriptions of blame: The ordinary ethics of farming communities. *Journal of Rural Studies*, *62*, 62–67.

Burton, R. J. F. (2004). Seeing through the 'good farmer's' eyes: Towards developing an understanding of the social symbolic value of 'productivist' behaviour. *Sociologia Ruralis*, *44*(2), 195–215.

Burton, R. J. F., Kuczera, C., and Schwarz, G. (2008). Exploring farmers' cultural resistance to voluntary agri-environmental schemes. *Sociologia Ruralis*, *48*(1), 16–37.

Burton, R. J. F., and Paragahawewa, U. H. (2011). Creating culturally sustainable agri-environmental schemes. *Journal of Rural Studies*, *27*(1), 95–104.

Burton, R. J. F., and Wilson, G. A. (2006). Injecting social psychology theory into conceptualisations of agricultural agency: Towards a post-productivist farmer self-identity? *Journal of Rural Studies*, *22*(1), 95–115.

Campbell, H. (2020). *Farming inside invisible worlds: Modernist agriculture and its consequences*. New York: Bloomsbury.

Carolan, M. (2020). Ethical eating as experienced by consumers and producers: When good food meets good farmers. *Journal of Consumer Culture*. [Online first]

Carolan, M. S. (2011). *Embodied Food Politics*. Abingdon, UK and New York: Ashgate Pub Co.

Coughenour, C. M. (1976). A theory of instrumental activity and farm enterprise commitment applied to woolgrowing in Australia. *Rural Sociology*, *41*(1), 76–98.

Duram, L. A. (1997). A pragmatic study of conventional and alternative farmers in Colorado. *The Professional Geographer*, *49*(2), 202–213.

Foot, P. (1969). Goodness and choice. In W. D. Hudson (Ed.), *The is-ought question* (pp. 214–227). London: Palgrave Macmillan.

Forney, J., Rosin, C., and Campbell, H. (2018). *Agri-environmental governance as an assemblage: Multiplicity, power, and transformation*. Abingdon, UK and New York: Routledge.

Gasson, R. (1973). Goals and values of farmers. *Journal of Agricultural Economics*, *24*(3), 521–542.

Haggerty, J., Campbell, H., and Morris, C. (2009). Keeping the stress off the sheep? Agricultural intensification, neoliberalism, and "Good" farming. *Geoforum*, *40*(5), 767–777.

Harrison, C. M., Burgess, J., and Clark, J. (1998). Discounted knowledges: Farmers and residents understandings of nature conservation goals and policies. *Journal of Environmental Management*, *54*(4), 305–320.

Johnsen, S. (2004). The redefinition of family farming: Agricultural restructuring and farm adjustment in Waihemo, New Zealand. *Journal of Rural Studies, 20*(4), 419–432.

Kessler, A., Parkins, J. R., and Huddart Kennedy, E. (2016). Environmental harm and "the Good Farmer": Conceptualizing discourses of environmental sustainability in the beef industry. *Rural Sociology, 81*(2), 172–193.

Lowe, P., Clark, J., Seymour, S., and Ward, N. (1997). *Moralizing the environment: Countryside change, farming and pollution*. London: UCL Press Limited.

MacIntyre, A. (1984). *After virtue*. Vol. 211. Notre Dame: University of Notre Dame Press.

Morris, C. (2006). Negotiating the boundary between state-led and farmer approaches to knowing nature: An analysis of UK agri-environment schemes. *Geoforum, 37*(1), 113–127.

Morris, C., and Evans, N. (2004). Agricultural turns, geographical turns: Retrospect and prospect. *Journal of Rural Studies, 20*(1), 95–111.

Murdoch, I. (2001). *The sovereignty of good*. Abingdon, UK and New York: Routledge.

Naylor, R., Hamilton-Webb, A., Little, R., and Maye, D. (2018). The 'Good Farmer': Farmer identities and the control of exotic livestock disease in England: Farmer identities: Exotic disease control. *Sociologia Ruralis, 58*(1), 3–19.

Ploeg, J. D. van der. (1994). Styles of farming: An introductory note on concepts and methodology. In J. D. van der Ploeg and A. Long (Eds.), *Born from within. Practice and perspectives of endogenous rural development* (pp. 7–31). Van Assen: Gorcum.

Riley, M. (2016a). How does longer term participation in agri-environment schemes [re] shape farmers' environmental dispositions and identities? *Land Use Policy, 52*, 62–75.

Riley, M. (2016b). Still being the 'good farmer':(non-) retirement and the preservation of farming identities in older age. *Sociologia Ruralis, 56*(1), 96–115.

Saunders, F. P. (2016). Complex shades of green: Gradually changing notions of the 'Good Farmer' in a Swedish context: The 'good farmer' in a Swedish context. *Sociologia Ruralis, 56*(3), 391–407.

Shortall, O., Sutherland, L.-A., Ruston, A., and Kaler, J. (2018). True cowmen and commercial farmers: Exploring vets' and dairy farmers' contrasting views of 'Good Farming' in relation to biosecurity: True cowmen and commercial farmers. *Sociologia Ruralis, 58*(3), 583–603.

Silvasti, T. (2003). The cultural model of "the good farmer" and the environmental question in Finland. *Agriculture and Human Values, 20*(2), 143–150.

Stock, P. V. (2007). 'Good Farmers' as reflexive producers: An examination of family organic farmers in the US Midwest. *Sociologia Ruralis, 47*(2), 83–102.

Stock, P. V. (2015). Contradictions in hope and care. In *Food Utopias: Reimagining citizenship, ethics and community* (pp. 171–194). Abingdon, UK and New York: Routledge and Earthscan.

Stock, P. V., Forney, J., Emery, S. B., and Wittman, H. (2014). Neoliberal natures on the farm: Farmer autonomy and cooperation in comparative perspective. *Journal of Rural Studies, 36*, 411–422.

Stuart, D., and Houser, M. (2018). Producing compliant polluters: Seed companies and nitrogen fertilizer application in U.S. corn agriculture: Producing compliant polluters. *Rural Sociology, 83*(4), 857–881.

Stuart, D., Schewe, R. L., and McDermott, M. (2012). Responding to climate change: Barriers to reflexive modernization in US agriculture. *Organization & Environment, 25*(3), 308–327.

Sutherland, L.-A. (2013). Can organic farmers be 'good farmers'? Adding the 'taste of necessity' to the conventionalization debate. *Agriculture and Human Values, 30*, 429–444.

Sutherland, L.-A., and Darnhofer, I. (2012). Of organic farmers and "good farmers": Changing habitus in rural England. *Journal of Rural Studies, 28*(3), 232–240.

Tall, I., and Campbell, H. (2018). The "dirty dairying" campaign in New Zealand: Constructing problems and assembling responses. In *Agri-environmental governance as an assemblage* (pp. 161–176). Abingdon, UK and New York: Routledge.

Tsouvalis, J., Seymour, S., and Watkins, C. (2000). Exploring knowledge-cultures: Precision farming, yield mapping, and the expert – farmer interface. *Environment and Planning A: Economy and Space, 32*(5), 909–924.

Williams, D., Lloyd, T., and Watkins, C. (1994). *Farmers not foresters: Constraints on the planting of new farm woodland.* Working Paper 27, University of Nottingham.

Wuthnow, R. (1989). *Meaning and moral order: Explorations in cultural analysis.* Berkeley, CA: University of California Press.

6

THE GENDERED 'GOOD FARMER'

This chapter addresses the gendered nature of the good farmer concept, and in particular what it means for women to be understood as a 'good farmer'. In Chapter 2 we observed how, in the 16th and 17th centuries, the 'good wife' was an essential part of men being recognised as a good farmer/husbandman. Two reasons for this could be discerned. Firstly, the wife performed essential economic roles on the farm such as the making of linen and cheese and the breeding of livestock, as well as managing and directing the labour. Secondly, as the definition of a good farmer/husbandman incorporated moral aspects to a greater degree, the good farmer had to be a good family man, and to demonstrate he was a good husband, he needed to be able to show that he had a 'good wife'. As time progressed, the moral imperative of the farmer became production rather than God and family, while the essential role of the farm wife as a manager and manufacturer diminished. The role of the farmer and wife became strongly gendered.

This gendered role for the farm wife continues to be a significant part of contemporary agriculture. In the global North, farmers are still presumed to be men (Shortall, 2016) while in many parts of the world men are culturally presumed to have a professional career, of which farmer is one possibility. In contrast, if women have a profession it is in addition to their role as a woman, a homemaker, a wife, or some other social role. As Beach (2013) notes:

> The label 'farmer' is usually associated with masculine traits and men. . . . While men are often identified in relation to their work . . . women on farms have been defined in terms of their marital relationships as farmwives instead of in terms of their connections to the land, the farms, or their children as farmers, farm women, or farm mothers. The farmer's wife is portrayed as the supporter of the male farmer, the homemaker, the one who raises the children and cares for the elders.
>
> *(pp. 212–213)*

Without avenues for professional success within agriculture, women have been made invisible as farmers and thus made to be something other than a farmer. In many ways this situation parallels the invisibilising of women under patriarchy. In this chapter, we both highlight the effects of these exclusions and seek opportunities to expand research to actively look for other examples of good farmers that have been similarly marginalised.

To be a good farmer in many contexts now presumes the farmer to be a man. For the most part, studies on the 'good farmer' have followed a phenomenological line of inquiry, thus helping maintain the role of the wife as 'helper'. As such, those tasks thought of as farming (e.g. planting, harvesting, showing and breeding livestock, making contracts) are generally assumed to be male activities, while home-based organisational activities such as running errands, household management, and accounting tend to be thought of as female (Woods, 2011, p. 218). The male activities evolved metrics of success, often in parallel with systems of professionalisation. Ever-increasing professionalisation developed as a visible symbol of success whereas women's contributions remained local, hidden, vague, and unmeasurable. This leads to tensions and contradictions when claiming the mantle of 'farmer' that are wrapped up in land tenure, marital status, education, personality, geography, type of operation(s), the presence of children, along with other variables (Sachs et al., 2016).

Women's work on farms is clearly not limited to 'helping' or reproductive labour – women on farms participate in the full range of farm activities (Shortall et al., 2017), particularly, animal husbandry, accounting, "running errands, supervising farm labour, growing and preserving food, various farm chores, and 'filling-in' or doing what 'needs to be done'" (Sachs, 2018, p. 134). Women also lead farms in their own right: According to the 2012 United States Census of Agriculture, of 2.1 million farmers or 86 percent of "primary operating" (i.e. primary decision-maker) famers were male, and 14 percent were female (Hoppe and Korb, 2013). The figures are similar in the United Kingdom, where in England, 84% of farm holders in 2016 were male. Across Europe, percentages of female-led farms range from 4 to 45% (Eurostat, 2019). With the exception of countries like Lithuania and Latvia, where the percentage of female-led farms approaches 50%, the number of female farmers is considerably lower than the percentage of women identified as working in the agricultural sector.[1]

The academic literature also demonstrates that although the position of women in society has changed in recent decades (e.g. increased labour force participation, higher levels of education), farming as a sector remains male dominated. Mechanisation – which should negate any biological differences in the ability of women to carry out farm work – has instead led to further polarisation of gender roles (Heggem, 2014; Saugeres, 2002b). Efforts to rebalance inheritance of agricultural land have been largely unsuccessful (Shortall, 2010). Gender disaggregated data about farm work and rural governance structures remains elusive (Shortall, 2016), although it is recognised that when women lead farms, those farms tend to be smaller and less commercial in orientation (Sutherland et al., 2016). Smyth et al. (2018, p. 671) contend that whereas "women seem to be adapting and changing;

the institution of farming and ranching does not". Agriculture across the global North continues to be patriarchal.

While much of the good farmer literature has focused on the phenomenological experiences of farmers (what meaning being a farmer gives their lives), the majority of those experiences are those of male farmers. In this chapter, we use that reality as a starting point to, first, acknowledge this weakness in the good farmer literature and explore the production of the 'good farmer' identity from a gender perspective. Second, we address the key questions of how good farmer identities are gendered and whether the good farmer construct reinforces traditional gender roles and the 'invisibility' of women in agricultural adjustment research. Third, we take an historical look at some symbols related to identification of a woman as a good farm wife. Fourth and finally, we conclude with sections on excluded others and suggestions for areas of dialogue that studies of the good farmer might more thoroughly engage with in the future. Significantly, we ask, can systems of farm professionalisation emerge as change agents for inclusion and equity?

Hiding gender in the good farmer literature

In Chapter 1 we observed how studies of women's roles and identities in agriculture preceded the good farmer identity concept by more than a decade (as noted in Burton, 1998), yet this connection is barely acknowledged in contemporary studies, let alone explored. In fact, a review of the good farmer literature to date demonstrates the that the good farmer concept, as with the economic and attitudinal studies before it, also defaults towards male perceptions. Although women are sometimes identified as study participants (particularly in studies addressing the good farmer in relation to 'alternative' forms of agricultural production), the concept of the 'good farmer' is rarely explicitly applied in relation to the role of women on farms. For example, Burton's (2004) and Burton et al.'s (2008) papers on the 'good farmer' make no mention of gender;[2] respondents are identified solely by number. Studies of the good farmer such as those of Stock (2007), Hunt (2010), Sutherland and Darnhofer (2012), Sutherland (2013), Saunders (2016), Shortall et al. (2018) and Naylor et al. (2018) include a few female participants in their sample, but do not address gender dynamics.

These studies are characteristic of recent sampling practices within the rural studies and agricultural adjustment literatures, where researchers typically engage with the 'primary farmer' – i.e. one qualitative interview or survey per farm, in some cases a joint interview with a farming couple, but primarily with individual men. Women are thus typically interviewed as part of a couple, if they are interviewed at all. As Sarah Whatmore pointed out in her book *Farming Women* (1991, p. 4), "Analytical attention has focused on the 'farmer' as business principal, labourer, and decision maker (Errington et al., 1986), the term itself carrying masculine connotations". For the most part, the good farmer literature tracks along a similar path because this literature began as a way for farmers (mostly men) to describe the status of being good in each other's (again, primarily men's) eyes.

Visual symbols of the good farmer foreground men and a specific kind of rural and agricultural masculinity whereby the symbols of being a good farmer also tend to reflect being a good man (Silvasti, 2003). As discussed in Chapter 3, symbols of the 'good farmer' typically include even, weed-free fields, high quality, healthy livestock, and a tidy farmstead. These symbols also reflect a financially viable farm – profits are expected from quality crops and livestock; tidy farmsteads indicate that capital and labour is available for maintenance (Sutherland, 2013), thus reinforcing the role of male as both professional and provider. Men typically drive the tractors (i.e. demonstrate the skilled role performance that leads to weed-free fields) (Heggem, 2014). Men are also dominant in the major visual spaces of agriculture, such as auction marts (Pilgeram, 2007). Riley (2016), in focusing specifically on masculinity and the good farmer, argues that it is important to farmers' identity as they age that they maintain a visible, public presence as a farmer, as well as the financial viability of the farm. Farmers are thus typically reluctant to shift into full, formal retirement. Riley concludes by recognising that further research is needed to assess the intersecting roles of other farm household members including spouses.

The lack of attention to women in the good farmer literature raises a number of issues. First, as farm decision-making is typically undertaken at the household level (Farmar-Bowers and Lane, 2009; Gasson and Errington, 1993), interviewing a single (male) household member excludes or marginalises the experiences and actions of both women and young people (e.g. successors), thus contributing to making women 'invisible' (Haney and Knowles, 1988; Riley, 2009). Second, by avoiding gender debates, these studies appear gender neutral – whereas in this chapter we question whether this is actually the case. Third, as men form the vast majority of study participants, good farmer studies promote the image of farming as predominantly male. Fourth, these studies may also disproportionately represent male perspectives.

Although empirical work on the good farmer largely ignores dynamics between the genders, conceptualisation of the 'good farmer' can provide useful insights into how gendered roles on farms have been reproduced in a Bourdieusian sense (described in Chapter 4). Saugeres (2002b) in her French study of mechanisation in farming, argued that the discourse surrounding the role of women on farms revolved around what is 'natural' for women – that women are less biologically suited to both heavy labour and operating machinery. Saugeres (2002a) also contended that men, as inheritors of land, become embodied in the land and the acts of cultivation. This 'naturalness' is reinforced by women who marry into farming who may not have tractor-driving skills and can be intimidated by large (expensive) machinery. In contrast, the gendered socialisation of care-work that is central to the habitus of many women can reinforce inequitable forms of familial organisation. DeVault (1991) asserts that this makes it seem as if caring is a natural female trait. As a result, generations of women have been oriented to unpaid and low paid care-work. Women's labour is also primarily indoors – inside the house, or inside the barn, where women have historically taken responsibility for animal husbandry (Alston, 1995). Work in the fields was oriented towards 'helping' (e.g. during busy periods such as harvest). Thus, studies adopting the good farmer concept are primed to make observations that de-naturalise gender-based assumptions about work related to farms.

The perpetuation of these gender norms is reinforced by farming families. Multiple family members are involved in socialisation processes. Socialisation in farm households is undertaken by both male and female parents. Children are socialised by parents and peers to develop specific skills, including the ability to produce and recognise culturally valued symbols. Men and women on farms reinforce the continuation of established gender roles. Fischer and Burton (2014) argue that socialisation processes in childhood shape the subsequent career choices of farm children – boys identified as the successor are much more actively engaged in developing farming skills as children which, consequently, reinforces their successor identity and teaches them what it is to be a 'good farmer'. Implicit in the paper (and explicit in Fischer, 2007) is that girls raised on farms are not typically identified as successors, or given the opportunity to learn these skills and, as a consequence, do not develop farmer or successor identities. Brandth (2002) argues that this perpetuation of gender roles is continued even in cases where women earn substantial off-farm income. Rather than changing the relative social position of women in the household, farmers' wives actively seek to promote the husband's identity as provider and farmer, as a means of supporting his mental health (Shortall, 2014). As a result,

> It is the man who typically claims the title 'farmer,' even on a family farm where the 'farm wife' and the 'farm kids' labour both in the fields and in the home on tasks essential to the farm enterprise: feeding livestock, driving grain wagons to the elevator, balancing the books, washing the clothes and dishes, cooking the food, and acting as reserve drivers. . . . [E]very farm typically has only one 'farmer.'
>
> *(Campbell et al., 2006, p. 5)*

Women themselves may be uncomfortable with claiming the identity of farmer. Michael Bell's (2004) study of farmers and farms transitioning to sustainable production systems with the attendant changes in identity, found women reluctant to claim the mantle of 'farmer' – with one widow failing to do so even after being the sole farmer for ten years. Traditionally, widows have been understood as placeholders, maintaining the farm until a male relative can take over (Carter, 2017). Writing about a somewhat conservative district in New Zealand, Hatch (1992) noted:

> A woman may own a farm by herself, and conceivably she could do all the work on it that a male owner would do . . . but even so, to refer to her as a farmer would strain normal usage to the limit. The business of farming is conceived as inherently masculine, and this is so even though women's labour is extremely important on many properties. . . . A woman may be highly regarded for her contribution to the farm operation, yet the labour

> she provides is considered supportive, not primary, and in an important sense she is viewed as subordinate to her husband in the operation of the business.
> *(p. 11)*

Rosenfeld (1985) describes how, despite the fact that women do a significant amount of farm work, only 4% of women claimed the title of farmer on tax forms in (at the time) the only nationally representative survey of farm women and work in the United States. The inability or unwillingness to claim the title of farmer for many women then leads to material and financial consequences; if for no other reason than because others, such as financial institutions, do not accept that women can be farmers (Contzen and Forney, 2017; Droz et al , 2014; Stock et al., 2019).[3]

In the sole 'good farmer' study of women on farms to date, Heggem (2014) argues that traditional gender perspectives in agriculture emphasize the gendered nature of farm roles as a matter of nature over nurture. Respondents (both male and female) interviewed in her Norwegian study observed how they believed male interest in machinery was due to the 'tractor gene' – a male disposition that predetermined an interest in machinery and other traditionally male roles in agriculture, interests that women were simply not born with. Fischer (2007) and Heggem (2014) both produce compelling arguments against this perspective, arguing that the interest in machinery is enculturated from childhood through gendered engagement with agriculture rather than being part of a predetermined gender preference. Heggem (2014) further contends that, in the Norwegian context, increasing engagement in tourism and Green Care is reshaping primary activities on the farm and leading to new symbols of the 'good farmer'. These activities have characteristics that are typically more attractive to women as they are more compatible with women's educational choices and, consequently, with women taking over key roles on the farm they are now more likely to be identified as suitable successors. This is explored later in the chapter. First, we consider the historic cultural symbols of women on farms as antecedents to contemporary imagery.

Symbols of the Good Farm Woman

Without opportunities to produce professionally articulated symbols of being a successful farm woman, women have negotiated their own symbols of what it means to be a good farm wife. How then do women represent their contribution to the success of the farm to other women and their communities? Lauters' (2009, p. 80) cultural history of farm women utilised farm periodicals in the United States from 1910 to 1960 to demonstrate that farm women were portrayed, "as the architects of farm life and keepers of the farm home and as important and central to the business of farming". In line with this, O'Hara (1998) identified four categories of farm women's labour relationships: farm helper, farm homemaker, farmwives 'working for the family farm', and farm women in paid work (where the work is not on the farm). While following the basic categorisation of O'Hara, we contend that it is in effect contextually dependent, such that in a different set of circumstances the symbols of group belonging are likely to also differ. Acknowledgement of these categories is also likely to vary. For example, in some cases the laws were such that women could not even be considered as legal title holders – meaning that they were technically unable to assume the full role of the farmer by having full control over the land (Keller, 2014).

In this section we consider the historic symbols of the good farm woman. Wilkie (2010) describes the three historic zones of farm labour: the field, farmyard, and house. Traditionally in the United States, field production (of crops) was primarily undertaken by men and household labour primarily by women, with the two meeting in the farmyard. "The primary ingredient for success was hard work" (Lauters, 2009, p. 90). Farm women demonstrated hard work in five primary areas: animal care, 'domestic manufactures' or cottage industries, (forced) hospitality, incorporation of mechanisation, and civic engagement. Whilst the following sections emphasize the US situation, most elements echo general trends that have been documented in other Western contexts, particularly the United Kingdom.

Mechanisation

Over the course of the 20th century, mechanisation of both farm and household machinery altered farm work for both men and women. Jellison (1993) observed that circa 1930 in Shelby County, Iowa,

> farm women . . . periodically worked in the farm fields, particularly in corn picking and haying seasons. In deference to traditional notions about appropriate gender roles, however, women characterized such labour not as 'real work' but as 'helping' their husbands.
>
> *(p. 109)*

Household labour changed in response to the addition of household vehicles and other 'labour saving devices' as is evident, for example, in the scaling up of poultry and egg production to rely on more mechanised systems. An 1891 article in the *Farmer's Wife* newspaper (McCormick, 1891) notes,

> New machinery has done much to lighten and lessen the work of men on the farm – riding plows, patent drills, self-binders, headers and steam threshers all tend to do this; but machinery has yet wrought but little benefit to farmers' wives.
>
> *(p. 1)*

Many of the disagreements on farms in the early part of the 20th century were not between whether one should or should not mechanise, but whether the farm machinery or household machinery should be the first investment (Harper, 2001; Jellison, 1993). Harper's (2001) *Changing Works* documents how frustrated farm wives were as a result of the disparity between mechanised adoption of farm equipment and home mechanisation in upstate New York. Harper (2001, p. 193) compares "portrait[s] of patriarchy mitigated by gender flexibility inspired by evolving technology". Or, as Susan Rogers' (1975) work inspired Lauters (2009, p. 32) to state, "Balance is maintained within [farming] communities between the informal power of the women and the overt power of the men". In the United Kingdom, Riley (2009) documents women's pride in promoting a shift towards mechanisation on the farm that gave them more autonomy. This move was particularly evident during World War II, when the push to encourage women into field production was tied directly to the winning of the war. After the war, 'Tractorettes' were expected to move back into the house and revive the 'domestic ideal' (Jellison, 1993).

Brandth (1994) argued that the masculinisation and professionalisation of farming became particularly marked after the mechanisation of agriculture: machines, as associated with physical strength and mechanical skills, were seen as masculine. Men alone are featured in tractor advertisements, alongside text emphasizing the power, precision, and control offered by the machines. Tractors, and men, are pictured as large and strong, dominating nature. Saugeres (2002b) describes how mechanisation has reduced the visibility of women's work by limiting the presence of women in publicly visible fields – women are seen as less naturally disposed to driving tractors, and the tractor itself has become a symbol of male power. Instead of making it easier for women to undertake farm work, tractors are seen as making women's work in the fields unnecessary. Here we have the co-development of technology and the related professionalisation of farming acting together to literally hide women in the house. The gender relations hiccup in farming of World War II offers an interesting example of how women were able to contribute to farm success in the same way as men, but how the strength of gender stereotypes acted to draw them back into traditional roles after the war. Symbols of women's successful contributions to the farm were thus, once again, forced to take other forms.

Animals

Prior to World War II, in both the United Kingdom and the United States, women on diversified farms were typically responsible for dairy and poultry production (Lauters, 2009, p. 142; Wilkie, 2010, p. 48). Harper (2001, p. 7) chronicles the "transformation of farming during and just after World War II" and notes:

> The income from the eggs and poultry was 'women's money' (often denigrated as 'pin money'), bartered for the groceries needed to supplement self-provisioning and also traded and sold to pay for services such as doctoring and even, in some instances, new land and equipment.
>
> *(p. 186)*

Indeed, although income from eggs could exceed income from the dairy herd, poultry were regarded as a nuisance rather than a part of the farm system or the family economy (Harper, 2001; Wilkie, 2010). During and after World War II the demand for eggs grew such that policies shifted to encourage larger scale production on specialised farms, a move that made it somewhat more accepted as men's work – though not completely (Jellison, 1993, pp. 156–160).

Industrialisation in the dairy sector also served to 'de-feminise' milk production. Women were increasingly replaced by male workers in US and UK dairies, but strong norms of milking as women's work dissuaded men from milking roles in some countries (e.g. Sweden – Sommestad and McMurry, 1998). Women are also described in the literature as being responsible for caring for the calves, often including undertaking the calving process itself. As one woman put it, women were always taking care of the calves: "Because we treat 'em like babies. And they grow better, you know if you give 'em that little extra attention and fuss over 'em a little bit and tell 'em they look nice!" (Harper, 2001, p. 198). Once the calves were older, they became the responsibility of male staff. There is a gendered hierarchy of animal care on farms – men tend to be associated with the larger, more prestigious animals (mostly cattle), whereas women are associated with smaller livestock such as poultry and sheep (Sachs, 1996). Men also tend to be more visible where livestock are on display, such as action marts and agricultural fairs (Pilgeram, 2007). Thus, gender power dynamics play out through the symbols of successful animal rearing and breeding.

Domestic manufactures

In the 16th century, as noted in Chapter 2, the role of the farm wife in manufacturing goods from products grown on the farm made her of critical importance to recognition as a 'good farmer'. This role has continued in more recent times, although as manufacturing has moved into the cities, so the role of the farm wife as a manufacturer of everything from candles to linen has declined. Where domestic manufacturing has been able to – and is still able to – demonstrate professional status for farm women is in the arena of agricultural fairs where prowess can be demonstrated through what Neely (1935, p. 63) termed "domestic manufactures". As Smith (2011), writing about the Iowa State Fair, observes:

> Women have claimed spaces within the State Fair to make their own interventions in public culture and to influence the social, economic, and physical landscapes of Iowa. This gendered work is not a new phenomenon but grows directly out of rural work patterns of the late nineteenth and early twentieth century. While women's work was typically bound by gender expectations, it was also valued as essential to the farm economy and the larger farm community system.
>
> *(p. 25)*

Quilting for household use, exchange, and sale was a particularly important 'neighbouring' activity performed by groups of women (Smith, 2011). As a social activity, quilts were subject to peer assessment and emerged as an indicator of skill and dedication to the task in the same way that tending fields acted as symbols for male farmers. The symbolic importance of quilting could also extend beyond the local quilting circle. Endorsed as a farm-activity, the country fair provided "a physical domain in which participants represent their lives to themselves and all other interested parties" – both for the farmer through agriculture and the farm wife through activities such as quilting (Prosterman, 1995, p. 15). Participation at county fairs extended beyond quilts to include the production of food products, linens, artwork, and other 'domestic products'. However, as was the case centuries before, improvements in industrial manufacturing and processing combined with increasingly globalised trade saw the importance of these practices gradually diminish. As products became available for purchase at a significantly lower price than was possible with hand manufacture, the manufacturing role of farm women became increasingly economically insignificant (Rosenfeld, 1985).

Other cottage industries also contributed to the farm enterprise as revenue generators. Over time, these grew to include items that did not consist solely of things manufactured on the farm and marketed by farm women – but developed as businesses in their own right in order to support the farm enterprise. A similar development of women's engagement with the farm can be seen in contemporary literature on farm diversification. Little (2016) argues that entrepreneurship is typically understood in 'masculine' terms (power, control, and competitiveness) but, judging by the fact that the female production role has been maintained throughout the centuries, entrepreneurship is as much – if not more – a female role as a male one. However, there are some differences between male farm diversification and female farm diversification. In a Dutch study, Bock (2004) demonstrates that when women develop diversification activities on farm, they tend to do so at a smaller scale, and are oriented towards fitting the new activities around their existing tasks, particularly childcare, in order to retain their status as good mothers and farm wives. Further, these diversification activities were intended to supplement farm income rather than accumulate capital and were criticised for undermining male achievements (i.e. undertaking diversification suggested that the farm was not economically viable in itself).

Hospitality

In various parts of the world, farmers hire seasonal labour. Harper's (2001) description of the changing system of shared labour and equipment related to grain growing and dairy farming suggests that many of the hired men could expect room and board as part their labour agreement – and it would generally fall to the farm wife to provide this. For the most part her participation was expected and not negotiated (Jellison, 1993). Some enjoyed it. Others described the "irritation" of integrating hired help into the family – doing the washing, cooking, and additional care-work (Harper, 2001, p. 96). Edith Bradley Rendleman who lived on farms in southern Illinois in the first half of the 20th century observed "That is one thing I hated, to always have somebody around to cook, wash, and iron for. You never got to enjoy your own family alone" (quoted in Harper, 2001, p. 96). As Whatmore (1991, p. 95) contends, this kind of hospitality expectation, illustrates the " 'blurring' of the traditional division between 'domestic' and 'social labour' on family farms". The twin developments of the professionalisation of farming and the nuclear family created strain around the family dinner table.

This echoes more recent observations on the consequences of farm diversification for women. Recent research has suggested that farm women are generally given (or adopt) the role of managing diversification activities where public interaction is required (e.g. direct sales, agri-tourism) (Wright and Annes, 2014). These interactions are understood to be more within women's skill sets and 'natural orientations'. Brandth and Haugen (2010, 2011), for example, describe how when farming couples diversify into tourism activities, gender and work identities are renegotiated. This may lead to traditionally feminine acts of hospitality becoming more highly valued within the farm business (Heggem, 2014), but can still reinforce patterned understandings of masculine and feminine roles.

Political activity

Lauters (2009, p. 15) contends that along with other areas of farm and social life, farm women are "active citizens in rural communities". Whereas much of the rural literature excludes farm women as active civic actors (outside of the family and church), Lauters' examination of periodicals, especially *The Farmer's Wife*, concludes that many farm women supported, at least within the confines of the periodical (through contributed letters), the suffragist movement, a wider voice for farm women, and the importance of their work – especially through the Great Depression. In one instance, Emma D. Pack (1892) wrote,

> It is the farmer and the working man that believe the women their equal. . . . Every man who loves his wife and children should commence to think of their future. Let the mothers vote. All people regardless of colour or sex would be benefitted by the reforms demanded by the Farmers' Alliance.
>
> *(p. 5)*

Sara Egge and Jenny Barker Devine's (2014) work on the same era of US farm women illustrates political involvement that helped to make the 'feminist' umbrella a little wider than many urban feminists might have imagined possible. Their work illustrates the murkiness of movement labels and categories that hinge on similar us-versus-them categories that parallel our good farmer constructions.

This murkiness carries over into the role that farm women's organisations play (Rosenfeld, 2017; Sachs, 2018). While mainstream agricultural organisations continue to serve as an avenue for knowledge sharing, they remain mostly male-centric both in terms of the knowledge shared and their leadership structure (Alston and

Wilkinson, 1998). Female-centred agricultural organisations privilege the unique experiences of women on farms as opposed to the presumed mainstream. Further, female leaders play a vital role in increasing both women's sense of being a farmer as part of their identity and shifting policy towards more gender equitable outcomes (McVay, 2016). This relationality adds nuance to the role that gender, marriage and parenthood, friendship, care, and dialogue play in the evolution of both the symbols and the practices of being a good farmer.

This section illustrated how the centrality of women's participation on the farm has been downplayed and how women have consequently been excluded from consideration as good farmers – except in unique circumstances. Historically, women's symbols of success, of being good farm women, have revolved around the creation of domestic objects such as quilts, raising smaller (thus feminine) animals, and contributing to the overall well-being of the home – welcoming guests, preparing meals, and limiting negative attention to the farm. These symbols have been attributed as naturally female – even when the lead farmer on a property was a woman.

The good farmer as an expression of masculinity

What it means to be a good farmer has also traditionally been related to the construction of masculinity in farming. For the good farmer to be inherently masculine also means that what is a 'good farmer' has been constructed in dialogue with a definition of rural masculinity. Charatsari and Istenič (2016, p. 392) contend that "in many studies male farmers are considered as a stable reference standard against which gender differences are evaluated often without being included in the study sample". In essence, male farmers are treated as a baseline against which women are compared, furthering the implicit position of men as 'real' farmers and women as 'other'. Contemporary research emphasizes that gender identity is relational: masculinities are heterogeneous and often influenced by changes in women's practices (Haugen and Brandth, 2017).

Barlett and Conger (2004) identify three visions of masculine success on farms: industrial, agrarian, and an emergent ecological vision. We address the first two in this section while taking up the third in the next section. In the *industrial ideal* the male is the provider and the farm the means of provision. Anything that indicates he is not able to provide for the family diminishes his masculine identity and, as such, the fact that his wife does not have to work on the farm becomes a symbol of success as both a man and a farmer (also see Barlett, 1993). Women thus become completely separated from the means of production. A 'good' (male) farmer is one who has "tamed the elements to produce crops and manage livestock, overcoming nature's vagaries and uncertainties" (Little, 2006, p. 189) – with the shift towards machinery and chemical inputs exhibiting masculine control over nature. To operate a large commercial business and complex machinery is to be professional, and that professionalism is invariably masculine (Bell et al., 2015).

Female farmers positioning themselves within this environment adopt stereotypically male symbols of success. For example, Pilgeram (2007) writes, that:

> Rather than the women's presence 'eroding the gender stereotypes that one has be "tough, strong, dominant, and male" to be a good farmer', the women used these stereotypes in order to present themselves as good farmers, even at the expense of their personal identity as women.
>
> *(p. 585)*

She details how female farmers at American auction marts distinguish themselves from farmers' wives at the same events by wearing masculine clothing, using rough language, and positioning themselves near to the auction ring, in contrast to farmers' wives, who adopt feminine symbols (e.g. wearing make-up and carrying handbags, locating themselves at a distance from the ring). Smyth et al. (2018) found:

> women's involvement in farms and ranches is related to their gender self-perception, with more involvement being associated with a more masculine self-perception, especially for involvement in manual labour jobs and jobs that involve horses. Women who view their primary role as independent agricultural producers or full partners also perceive themselves as more masculine than women who view their primary role as homemaker.
>
> *(pp. 669–670)*

Barlett and Conger's (2004) *agrarian ideal* of masculine success emphasizes yeomanship and the ability to perform farm work; farming is a family commitment and explicitly involves women. Examples in the 'good farmer' literature emphasize this agrarian perspective – the skilled role performance inherent in livestock husbandry and crop production, rather than the direct relationship between the farm and financial success. This may be because, as Burton (2004) observes, symbols of financial success are 'hidden' whereas skilled role performances in the field are visible, unmistakable, and directly attributable to a specific farm. Sutherland (2013), however, argues that increasing economic pressures are enhancing he importance of financial success (by whatever means) as a symbol of a good farmer. Bryant (1999) observes an association between what she terms 'feminine pride' and nurturing/helping roles in agrarian ideology, and contrasts these with the 'masculine pride' involved in undertaking physical labour. However, these ideas have yet to be developed more widely. For the good farmer literature it is important to avoid the same essentialising assumptions, i.e. that male 'good farmers' are more likely to engage in larger scale operations while female 'good farmers' are solely committed to smaller-scale and alternative productions.

The two identified types align with Contzen and Forney's (2017) typology of family farming in Switzerland. In this study, the majority of family farms fit with the agrarian vision, where the family and the farm are 'fully interdependent' in Switzerland as 'traditional' complementarity whereby "the origin of labour is defined by the clear separation of spheres of activity according to the classic gendered division between production and reproduction" (p. 33). Their results emphasize that, despite the still clear domination of a 'traditional' model, a diversification of farm configurations exists, whereby all individual roles are jointly renegotiated, offering new possibilities for gender equality. The industrial vision is markedly similar to Contzen and Forney's "professional individualism" configuration, whereby farm activity is kept as separate as possible from the household and family sphere, and the non-farming partner is not involved in the farm work (p. 34). For Contzen and Forney, however, this configuration also includes families where the woman has her own professional career and is not limited to her housewife role, as is suggested in Barrett and Conger's vision.

Failure to meet either the industrial or agrarian ideals can be particularly damaging to men's identity and mental health, e.g. depression(Valkonen and Hanninen, 2012), other mental illnesses (Parr et al., 2004), and emotional distress (Coen et al., 2013). Barlett and Conger (2004) argue that there are many examples of mental health issues or even suicide as a result of farmers failing to meet expectations for maintaining farm viability. Bell (2004) in visiting a white nationalist, militia group, surmises that many are

> 'dispossessed' farmers, those who used to farm, but lost the farm. The loss of the material farm also contributed to a phenomenological 'disorientation' whereby their opportunity to practice masculinity as they had learned it also disappeared. If you can't practice being a farmer, then your self is diminished.
> *(p. 135)*

Here, the gendered construction of what is it be a 'good farmer' accentuates – for male farmers – the feeling of failure, combining the professional and gender dimensions of identity. The need for off-farm employment can be seen as a failure in the role of men as provider (Barlett and Conger, 2004). Off-farm employment can also lead women to disengage from the farm, particularly if it has become a drain on family finances (Alston, 2017).

Gender and alterity: being a 'different' good farmer?

Barlett and Conger's (2004) third type of masculine success is an *emergent ecological vision*. In the ecological vision, the enjoyment of production and reduced use of inputs are championed. Women are equal partners in this relationship. The vision is aspirational – hope for change is associated with the emergence of alternative agricultural practices which challenge the norms of a range of farming trends. The emancipatory potential of alternative approaches is not solely for the ecological impacts of the dominant agricultural system, but the power relationships embedded in patriarchy and heteronormativity. The idea that an emergent ecological vision comes from a balanced relationship between men and women on heteronormative family farms can be extended beyond such parameters regarding who and what goes into being considered a good farmer. What it helps point to is that in varying times, places, and industries there exist competing notions of what constitutes a good farmer. When considering gender and sexuality, we are challenged to theorise who is a good farmer when not a professional, primarily white, male?

It is well recognised that women are more visible in alternative agri-food networks (AAFN) than conventional networks (Ilbery and Maye, 2015; Tregear, 2011). Contemporary AAFNs encompasses a broad umbrella including options for production (e.g. organic, low input), marketing (e.g. direct marketing, short food supply chains, fair trade initiatives), and interactions with consumers (e.g. community-supported agriculture, local food movements). Wilbur (2014) cites historical research by Schmitt (2006), which demonstrates the 'astonishing' number of female researchers who were involved in the establishment of alternative agricultural science (i.e. early biodynamic and organic agriculture). Organic farming in particular is more likely to involve female-led farms (Pedersen and Kjærgård, 2004).

Michael Bell's ethnography of the Practical Farmers of Iowa (PFI) offers a concurrent example of Barlett and Conger's (2004) emergent ecological vision that he describes as relational agriculture, which "represents a great increase in the participation and prominence of PFI women on their farms and in the organization itself" (Bell, 2004, p. 205). Bell further notes:

> Sustainable agriculture involves more than agricultural practice. It is about the practice of the self – or, better put, it is about the practice of selves, new selves, new men and new women and new dialogues between and among them. It is about new conceptions of the unfinished wholes within which we all live. It is about new cultivations of farming, and that means both new senses of what a farm is and can be and new senses of what a person is and can be.
>
> *(p. 203)*

Because conventional agriculture is rooted in family farm configurations, knowledge systems, and power relations that include specific gendered role models, alternative agricultural practices offer opportunities to renegotiate how gender is understood and performed at the same time as the farm practices are changed. Following this logic, Trauger (2004) argued that alternative agriculture gives women spaces to challenge accepted norms.

However, this hopeful proposition is also contradicted in the literature. For instance, Silvasti (2003, p. 146) argued that conventional farmers in Finland regarded organic farming as 'pretend farming' and as not 'real work' because they saw organic farmers as chasing subsidies. Women's higher visibility could therefore be attributed to the lower prestige associated with alternative agriculture. In addition, the physical practices of alternative agriculture do not necessarily lend themselves to gender parity. Wilbur (2014) drew attention to the traditional gender roles adopted by 'back-to-the-landers' in Italy, the physical demands associated with alternative agriculture (e.g. manual labour associated with low input production), and the need for considerable reproductive and domestic labour associated with accommodating WWOOFers, leads to a pattern of men working outdoors and women indoors. While he argues that the important issue for women in the study was the sense of freedom they gained by leaving professional urban jobs to go 'back-to-the-land', his female respondents also identified considerable angst at the discrepancy between the anticipated and practical realities of this transition. In an American study, Castellano (2015) posits that engaging in alternative agriculture is not empowering for women: it is instead a continuation of traditional gender roles surrounding food provisioning.

Sustainable agriculture also tends to presume a heteronormative family farm familial structure. Overall, in the mainstream agri-food literature, and beyond the discussion of women's role in farming, the issues of sexuality, race, and class tend to take a marginalised role in comparison to cultural and economic analyses (Leslie, 2017, 2019; Wypler, 2019). Charatsari and Istenič (2016, p. 389) "highlight under-researched area[s]" at the intersection of gender and farming with a focus on gay, lesbian, bi farmers. Leslie (2017, p. 748) challenges that, "investigating the sustainability of the family farm demands a critical examination of its embedded heteronormativity . . . as well as an exploration of alternative conceptions of farm families, homes, and workplaces" (see also Wypler, 2019). This leads to questions such as: How then do LGBTQ farmers navigate these extra layers of challenges to help build new alternatives to the dominant food system? Specifically, what symbols do they adopt in their farm practices that both help navigate what it means to be a (good) farmer while also helping challenge the status quo? What historical examples of sustainable, 'good farmers' do we have that do not reflect examples of professional, heteronormative success? How can the good farmer concept evolve to take in these historical challenges to both farm practices and knowledge exchange?

As the literature on intersectionality and the rural grows, we can look to some historical examples of women, especially lesbians, actively creating intentionally separate spheres of farm production that necessarily involved their own formulations of what it meant to be a good farmer on these, often, collective farm ventures. Though we do not have explicit 'good farmer' literature around these kinds of farms, Leslie (2019) offers that,

> Some queer farmers desired the more traditional combined sexual and business partnership. Others thought it would jeopardize their relationship and farm, so created queer cohabitational home spaces that intentionally redefined the family farm.
>
> *(p. 11)*

As Luis (2018, p. 130) writes, "Women's lands are unique in their production of a gendered difference that goes beyond the creation of space 'for women' to the creation of literally female space". As part of the rising number of intentional communities of the 1960s and 1970s in the United States, lesbian, rural collectives also grew. Documented in books like *Country Lesbians* (1976) (an Oregon lesbian separatist collective) and *Lesbian Land* (1985), the reality of these collectives offers challenges from the margins of the rural and who is farming. By necessity, then, these women's stories hold their own definitions of what it means to be a good farmer and the related symbols. Here, being a good farmer was less about professional achievements, but how much these accomplishments built or maintained the overall collective, separatist efforts. Recent studies aim to extend the work on masculinity and femininity focused specifically on the rural, sexuality, and farming (Gorman-Murray et al., 2013; Gray et al., 2016).

In reference to women on the land, Luis (2018, p. 95) writes, "Boundaries are essential to maintaining community in all of its nostalgic as well as its literal sense". When everyone is the same in a community, in Luis' case primarily white, lesbian women, "there is no way to work with the Other, for strangers are automatically excluded from the community" (p. 96). Specifically, queer farmers encounter additional boundaries to access to land, credit, and knowledge based on gender and sexual identities (Leslie, 2017; 2019; Wypler, 2019) that parallel historical barriers to Native American, Latinx, African American, and female farmers in the United Stated especially (Brewer and Stock, 2016). Leslie (2017) found, counterintuitively to them,

> most farmers did not encounter the overt heterosexism they expected. . . . Queer farmers themselves re-evaluated their preconceived notions of rural and agricultural heterosexism, as sustainable farming often facilitated building rural community relationships and ties across sexual and gender differences.
>
> *(pp. 765–766)*

LGBTQ farmers are often associated with smaller scale, community-oriented agriculture and emphasise different symbols to mitigate these injustices. Tactics such as the use of hetero-male tools like chainsaws and tractors and androgynous clothing help lessen perceived differences. For many males, this means adopting a more 'macho' stance (Leslie, 2017, p. 758) or as Leslie also emphasises, the often perceived gender-neutrality of farmer dress attracted some to the profession. Some heterosexual women, many queer women, and the trans farmers appreciated gender-neutral farmer dress. Casey, a vegetable farmer in her 20s, said, "I like the more androgynous aspect of farmer wear. . . . I like that I can wear pretty gender-neutral-type clothing." When dressing up for special occasions, Tracy liked how farmers simply put on a vest over their normal wear. Dianne, a semiretired farmer in her 60s, wore worn overalls, and said that it felt good to do a sit-down interview "dressed like this" (Leslie, 2017, p. 761).

Beyond dress, LGBTQ farmers anticipate that land ownership helps to legitimise their pursuit of the farming profession as the most visible form of symbolic power in agriculture. In this "quest for identity" (Annes and Redlin, 2013, p. 129) not just as LGBTQ farmers, but as (good) farmers, many celebrate the emergence of a wider queer path to the farming profession most notably in the developing power of the Cultivating Change Network (www.cultivatingchangefoundation.org/) that offers specific funding and professionalisation opportunities (Leslie, 2017; Luis, 2018).

The historical patriarchal dominance that good farmers could only be white, male landowners often pushed others (women, lesbians, African Americans) into separatist enclaves that sought self-sufficiency away from male-centred knowledge bases. As Nelly discussed in *Country Lesbians* (1976) about her experience of learning how to use a chainsaw,

> I have struggled to steal the knowledge of chain sawing from the male culture and to share it with my sisters. I called the forestry department and the chain saw stores for information many times without results. . . . Because I am a lesbian separatist I couldn't get first-hand information from male friends.
>
> *(pp. 108–109)*

The boundaries of how and where one can gain farming knowledge reflect the power injustice, but also serve as a mechanism to resist power structures. Learning how to chainsaw through respectful knowledge exchanges demonstrates an important symbolic assertion of autonomy and power by Nelly and others in similar situations. In recognition that having the right tools enables good farm practices to be carried out, companies like Green Heron Tools (www.greenherontools.com/) now make tools specifically designed for women.

Future directions

In this chapter we have drawn attention to the (lack of) treatment of gender in the good farmer literature. Statistically, the vast majority of 'primary farmers' in economically developed countries are identified as male, and when women engage in farming, they are most likely to do so as part of a heterosexual couple. It is perhaps unsurprising that the good farmer literature has focused on men to date, but in doing so it has contributed to the invisibility of women in agriculture and reinforced the practice of omitting gender dynamics from analyses of agrarian identity. The literature on women in agriculture remains distinct from other studies of agrarian transition – separate streams at conferences, specific journal articles addressing women, a subset of rural scholars focusing on gender, but few integrate it into 'mainstream' analysis. This situation reflects partly, often unconsciously, a dominant framing in the wider society of modern, professional, agriculture as a masculine world. In order to address this issue, we suggest that new methods and concepts are needed. The good farmer literature needs to go beyond its traditional reliance on one interview per farm and actively seek to engage multiple household members. Engagement with broader feminist literatures opens up opportunities to more closely examine issues of embodiment, enactment, performance of professionalisation, and more-than-human care relationships. Intersectionality is also relevant – not only to how the good farmer literature privileges male experiences, but heterosexual white male experiences.

Notes

1 The wide range of female-led farms in Europe also begs the question about whether 'leading' a farm is liberatory or drudgery. For example, the literature on repeasantisation strikes a celebratory tone of people entering farming or doing something entrepreneurial and collective in farming. In other situations, leading a farm indicates a peasant condition rife with poverty, inequitable access to markets, historical degradation, and a life one would rather escape.
2 Burton (1998) acknowledged failure to address the identity of the farmer's wife when looking at behavioural choice as a weakness in his study (later published in 2004), but argued, rightly or wrongly, that the normative influence of family members had already been internalised by the individual interviewed. In the 2008 study women were present in almost 50% of the interviews, yet this was not elaborated on in later publications.
3 Women farmers are also less likely to receive government subsidies and private loans for their agricultural endeavours. An estimated 43,000 female farmers have been denied more than $4.6 billion in farm loans and loan servicing from the U.S. Department of Agriculture alone (Hadlik 2012, https://womensenews.org/2012/09/grass-ceiling-overhangs-surge-in-female-farmers/). This lack of access to capital and agricultural education hinders growth potential and reduces the likelihood of female farmer survival (Trauger et al., 2008).

References

Alston, M. (1995). *Women on the land: The hidden heart of rural Australia.* Australia: UNSW Press.

Alston, M. (2017). The genderness of climate change, Australia. In B. Bock and S. Shortall (Eds.), *Gender and rural globalization. International perspectives on gender and rural development* (pp. 114–128). Wallingford, Oxfordshire: CABI.

Alston, M., and Wilkinson, J. (1998). Australian farm women – shut out or fenced in? The lack of women in agricultural leadership. *Sociologia Ruralis*, *38*(3), 391–408.

Annes, A., and Redlin, M. (2013). Conceptual and spatial migrations: Rural gay men's quest for identity. In A. Gorman-Murray, B. Pini, and L. Bryant (Eds.), *Sexuality, rurality, and geography* (pp. 129–141). Lanham and Plymouth: Lexington Books.

Barlett, P. F. (1993). *American dreams, rural realities: Family farms in crisis.* Chapel Hill: UNC Press Books.

Barlett, P. F., and Conger, K. J. (2004). Three visions of masculine success on American farms. *Men and Masculinities*, 7(2), 205–227.

Beach, S. S. (2013). "Tractorettes" or partners? Farmers' views on women in Kansas farming households. *Rural Sociology*, *78*(2), 210–228.

Bell, M. (2004). *Farming for us all: Practical agriculture and the cultivation of sustainability.* University Park: Penn State Press.

Bell, S. E., Hullinger, A., and Brislen, L. (2015). Manipulated masculinities: Agribusiness, deskilling, and the rise of the businessman-farmer in the United States. *Rural Sociology, 80*(3), 285–313.

Bock, B. (2004). Fitting in and multi-tasking: Dutch farm women's strategies in rural entrepreneurship. *Sociologia Ruralis, 44*(3), 245–260.

Brandth, B. (1994). Changing femininity. The social construction of women farmers in Norway. *Sociologia Ruralis, 34*(2–3), 127–149.

Brandth, B. (2002). Gender identity in European family farming: A literature review. *Sociologia Ruralis, 42*(3), 181–200.

Brandth, B., and Haugen, M. S. (2010). Doing farm tourism: The intertwining practices of gender and work. *Signs: Journal of Women in Culture and Society, 35*(2), 425–446.

Brandth, B., and Haugen, M. S. (2011). Farm diversification into tourism – Implications for social identity? *Journal of Rural Studies, 27*(1), 35–44.

Brewer, J., and Stock, P. (2016). Beyond extension: Strengthening the federally recognized tribal extension program (FRTEP). *Journal of Agriculture, Food Systems, and Community Development, 6*(3), 91–101.

Bryant, L. (1999). The detraditionalization of occupational identities in farming in south Australia. *Sociologia Ruralis, 39*(2), 236–261.

Burton, R. J. F. (1998). *The role of farmer self-identity in agricultural decision making in the Marston Vale Community Forest.* Unpublished PhD Thesis. DeMontfort University, Leicester.

Burton, R. J. F. (2004). Seeing through the 'good farmer's' eyes: Towards developing an understanding of the social symbolic value of 'productivist' behaviour. *Sociologia Ruralis, 44*(2), 195–215.

Burton, R. J. F., Kuczera, C., and Schwarz, G. (2008). Exploring farmers' cultural resistance to voluntary agri-environmental schemes. *Sociologia Ruralis, 48*(1), 16–37.

Campbell, H., Bell, M. M., and Finney, M. (2006). *Country boys: Masculinity and rural life.* University Park: Penn State Press.

Carter, A. (2017). Placeholders and changemakers: Women farmland owners navigating gendered expectations. *Rural Sociology, 82*(3), 499–523.

Castellano, R. L. S. (2015). Alternative food networks and food provisioning as a gendered act. *Agriculture and Human Values, 32*, 461–474.

Charatsari, C., and Cernic Istenic, M. (2016). Gender, farming and rural social research: A relationship in flux. In M. Shucksmith and D. L. Brown (Eds.), *International handbook of rural studies* (pp. 389–399). Abingdon, Oxon: Routledge.

Cheney, J. (Ed.). (1985). *Lesbian land.* Minneapolis, MN: Word Weavers.

Coen, S. E., Oliffe, J. L., Johnson, J. L., and Kelly, M. T. (2013). Looking for Mr. PG: Masculinities and men's depression in a northern resource-based Canadian community. *Health and Place, 21*, 94–101.

Contzen, S., and Forney, J. (2017). Family farming and gendered division of labour on the move: A typology of farming-family configurations. *Agriculture and Human Values, 34*(1), 27–40.

DeVault, M. L. (1991). *Feeding the family: The social organization of caring as gendered work.* Chicago and London: University of Chicago Press.

Droz, Y., Miéville-Ott, V., and Reysoo, F. (2014). L'agriculteur et la paysanne suisses: un couple inégal? *Revue suisse de sociologie, 40*(2), 237–257.

Egge, S., and Devine, J. B. (2014). Putting the community first: Feminism and rural American women's activism in the twentieth century. In B. Pini, B. Brandth, and J. Little (Eds.), *Feminisms and ruralities* (pp. 15–30). Lanham and Plymouth: Lexington Books.

Errington, A. (1986). The delegation of decisions on the farm. *Agricultural Systems, 19*, 299–317.

Eurostat (2019). Females in the field: More women managing farms across Europe. European Commission. *NEWS*, March 8. https://ec.europa.eu/info/news/queens-frontage-women-farming-2019-mar-08_en

Farmar-Bowers, Q., and Lane, R. (2009). Understanding farmers' strategic decision-making processes and the implications for biodiversity conservation policy. *Journal of Environmental Management, 90*(2), 1135–1144.

Fischer, H. (2007). *Succession on Scottish family farms: Socialisation processes and the construction of farmer identities*. University of Aberdeen (United Kingdom).

Fischer, H., and Burton, R. J. F. (2014). Understanding farm succession as socially constructed endogenous cycles. *Sociologia Ruralis, 54*(4), 417–438.

Gasson, R., and Errington, A. J. (1993). *The farm family business*. Wallingford, UK: Cab International.

Gorman-Murray, A., Pini, B., and Bryant, L. (Eds.). (2013). *Sexuality, rurality, and geography*. Lanham and Plymouth: Lexington Books.

Gray, M. L., Johnson, C. R., and Gilley, B. J. (2016). *Queering the countryside: New frontiers in rural queer studies*. New York and London: New York University Press.

Haney, W. G., and Knowles, J. B. (Eds.). (1988). *Women and farming: Changing roles, changing structures*. Boulder, CO: Westview Press.

Harper, D. (2001). *Changing works: Visions of a lost agriculture*. Chicago and London: University of Chicago Press.

Hatch, E. (1992). *Respectable lives: Social standing in rural New Zealand*. Berkeley and Los Angeles: University of California Press.

Haugen, M. S., and Brandth, B. (2017). The gendered embodiment of agricultural work: Nature, machinery and patriarchy. In *Gender, bodies and work* (pp. 99–110). Abingdon, UK and New York: Routledge.

Heggem, R. (2014). Diversification and re-feminisation of Norwegian farm properties. *Sociologia Ruralis, 54*(4), 439–459.

Hoppe, R. A., and Korb, P. (2013). Characteristics of women farm operators and their farms, EIB-111, U.S. Department of Agriculture, Economic Research Service, April.

Hunt, L. (2010). Interpreting orchardists' talk about their orchards: The good orchardists. *Agriculture and Human Values, 27*(4), 415–426.

Ilbery, B., and Maye, D. (2015). The changing dynamics of alternative agri-food networks: A European perspective. In G. Robinson and D. A. Carson (Eds.), *Handbook on the globalisation of agriculture* (pp. 425–445). Cheltenham, UK: Edward Elgar Pub. Ltd.

Jellison, K. (1993). *Entitled to power: Farm women and technology, 1913–1963*. Chapel Hill and London: University of North Carolina Press.

Keller, J. C. (2014). "I wanna have my own damn dairy farm!": Women farmers, legibility, and femininities in rural Wisconsin, US. *Journal of Rural Social Sciences, 29*(1), 75–102.

Lauters, A. M. (2009). *More than a farmer's wife: Voices of American farm women, 1910–1960*. Columbia and London: University of Missouri Press.

Leslie, I. S. (2017). Queer farmers: Sexuality and the transition to sustainable agriculture: Queer farmers. *Rural Sociology, 82*(4), 747–771.

Leslie, I. S. (2019). Queer farmland: Land access strategies for small-scale agriculture. *Society & Natural Resources, 32*(8), 928–946.

Little, Jo. (2006). Embodiment and rural masculinity. In H. Campbell, M. M. Bell, and M. Finney (Eds.), *Country boys: Masculinity and rural life* (pp. 183–202). University Park: Penn State Press.

Little, Jo. (2016). Gender and entrepreneurship. In M. Shucksmith and D. L. Brown (Eds.), *International handbook of rural studies* (pp. 357–366). Abingdon, Oxon: Routledge.

Luis, K. N. (2018). *Herlands: Exploring the women's land movement in the United States*. Minneapolis, MN: University of Minnesota Press.

McCormick, F. (1891). A Kansas farm. *The Farmers Wife*, *1*(3), 1.

McVay, L.-A. (2016). Leadership and gender. In M. Shucksmith and D. L. Brown (Eds.), *International handbook of rural studies* (pp. 367–378). Abingdon, Oxon: Routledge.

Naylor, R., Hamilton-Webb, A., Little, R., and Maye, D. (2018). The 'good farmer': Farmer identities and the control of exotic livestock disease in England. *Sociologia Ruralis*, *58*(1), 3–19. doi:10.1111/soru.12127

Neely, W. C. (1935). *The agricultural fair*. New York: Columbia University Press.

O'Hara, P. (1998). *Partners in production? Women, farm, and family in Ireland*. New York and Oxford: Berghahn Books.

Pack, E. D. (1892). Editorial. *The Farmers Wife*, *X*(7), 5.

Parr, H., Philo, C., and Burns, N. (2004). Social geographies of rural mental health: Experiencing inclusions and exclusions. *Transactions of the Institute of British Geographers*, *29*(4), 401–419.

Pedersen, K. B., and Kjærgård, B. (2004). Do we have room for shining eyes and cows as comrades? Gender perspectives on organic farming in Denmark. *Sociologia Ruralis*, *44*(4), 373–394.

Pilgeram, R. (2007). 'Ass-kicking' Women: Doing and undoing gender in a US livestock auction. *Gender, Work and Organization*, *14*(6), 572–595.

Prosterman, L. (1995). *Ordinary life, festival days: Aesthetics in the midwestern county fair*. Washington, DC: Smithsonian Institution Press.

Riley, M. (2009). 'The next link in the chain': Children, agri-cultural practices and the family farm. *Children's Geographies*, 7(3), 245–260.

Riley, M. (2016). Still being the 'Good Farmer': (Non-)retirement and the preservation of farming identities in older age. *Sociologia Ruralis*, *56*(1), 96–115.

Rogers, S. C. (1975). Female forms of power and the myth of male dominance: A model of female/male interaction in peasant society 1. *American Ethnologist*, *2*(4), 727–756.

Rosenfeld, R. A. (2017 [1985]). *Farm women: Work, farm, and family in the United States*. Chapel Hill: UNC Press Books.

Sachs, C. E. (2018 [1996]). *Gendered fields: Rural women, agriculture, and environment*. Boulder, CO: Westview Press.

Sachs, C. E., Barbercheck, M., Braiser, K., Kiernan, N. E., and Terman, A. R. (2016). *The rise of women farmers and sustainable agriculture*. Iowa City: University of Iowa Press.

Saugeres, L. (2002a). The cultural representation of the farming landscape: Masculinity, power and nature. *Journal of Rural Studies*, *18*(4), 373–384.

Saugeres, L. (2002b). Of tractors and men: Masculinity, technology and power in a French farming community. *Sociologia Ruralis*, *42*(2), 143–159.

Saunders, F. P. (2016). Complex shades of green: Gradually changing notions of the 'Good Farmer' in a Swedish context. *Sociologia Ruralis*, *56*(3), 391–407.

Schmitt, M. (2006). Fertile minds and friendly pens: Early women pioneers. *G. Holt and M. Reed: Sociological Perspectives on Organic Agriculture. From Pioneer to Policy*, 56–69.

Shortall, O., Sutherland, L. A., Ruston, A., and Kaler, J. (2018). True cowmen and commercial farmers: Exploring vets' and dairy farmers' contrasting views of 'Good Farming' in relation to biosecurity. *Sociologia Ruralis*, *58*, 583–603.

Shortall, S. (2010). *Women working on the farm: How to promote their contribution to the development of agriculture*. Retrieved from European Parliament. Directorate-general for internal policies. Agriculture and Rural Development.

Shortall, S. (2014). Farming, identity and well-being: Managing changing gender roles within Western European farm families. *Anthropological Notebooks*, *20*(3), 67–81.

Shortall, S. (2016). Changing configurations of gender and rural society: Future directions for research. In M. Shucksmith and D. L. Brown (Eds.), *International handbook of rural studies* (pp. 349–356). Abingdon, Oxon: Routledge.

Silvasti, T. (2003). The cultural model of "the good farmer" and the environmental question in Finland. *Agriculture and Human Values*, *20*(2), 143–150.

Smith, K. E. (2011). *Framing quilts/framing culture: Women's work and the politics of display*. Unpublished PhD Thesis. University of Iowa.

Smyth, J. D., Swendener, A., and Kazyak, E. (2018). Women's work? The relationship between farmwork and gender self-perception: Women's farmwork and gender self-perception. *Rural Sociology*, *83*(3), 654–676.

Sommestad, L., and McMurry, S. (1998). Farm daughters and industrialization: A comparative analysis of dairying in New York and Sweden, 1860–1920. *Journal of Women's History*, *10*(2), 137–164.

Stock, P. V. (2007). 'Good farmers' as reflexive producers: An examination of family organic farmers in the US Midwest. *Sociologia Ruralis*, *47*(2), 83–102.

Stock, P. V., Hossler, T., and Darby, D. B. (2019). Experiments in the field. *Visual and New Media Review, Fieldsights*, October 29. https://culanth.org/fieldsights/experiments-in-the-field

Sutherland, L.-A. (2013). Can organic farmers be 'good farmers'? Adding the 'taste of necessity' to the conventionalization debate. *Agriculture and Human Values*, *30*, 429–441.

Sutherland, L.-A., and Darnhofer, I. (2012). Of organic farmers and 'good farmers': Changing habitus in rural England. *Journal of Rural Studies*, *28*(3), 232–240.

Sutherland, L-A., Hopkins, J., Holstead, L., Toma, L., Barnes, A., and Matthews, K. (2016). *Female led-farmers in Scotland. Analysis of the 2013 CAP intentions survey. A report for RESAS*.

Trauger, A. (2004). 'Because they can do the work': Women farmers in sustainable agriculture in Pennsylvania, USA. *Gender, Place and Culture*, *11*(2), 289–307.

Trauger, A., Sachs, C., Barbercheck, M., Kiernan, N. E., Brasier, K., and Findeis, J. (2008). Agricultural education: Gender identity and knowledge exchange. *Journal of Rural Studies*, *24*(4), 432–439.

Tregear, A. (2011). Progressing knowledge in alternative and local food networks: Critical reflections and a research agenda. *Journal of Rural Studies*, *27*(4), 419–430.

Valkonen, J., and Hänninen, V. (2012). Narratives of masculinity and depression. *Men and Masculinities*, *16*(2), 160–180.

Whatmore, S., Lowe, P., and Marsden, T. (Eds.). (1991). *Rural enterprise: Shifting perspectives on small-scale production*. London: David Fulton Publishers Ltd.

Wilbur, A. (2013). Growing a radical ruralism: Back-to-the-land as practice and ideal. *Geography Compass*, 7(2), 149–160.

Wilbur, A. (2014). Back-to-the-house? Gender, domesticity and (dis)empowerment among back-to-the-land migrants in Northern Italy. *Journal of Rural Studies*, *35*, 1–10.

Wilkie, R. M. (2010). *Livestock/Deadstock. Working with farm animals from birth to slaughter*. Philadelphia, PA: Temple University Press.

Womanshare Collective (1976). *Country lesbians: The story of the Womanshare Collective*. Grants Pass, OR: Womanshare Books.

Woods, M. (2011). The local politics of the global countryside: Boosterism, aspirational ruralism and the contested reconstitution of Queenstcwn, New Zealand. *Geojournal*, *76*(4), 365–381.

Wright, W., and Annes, A. (2014). Farm women and agritourism: Representing a new rurality. *Sociologia Ruralis*, *54*(4), 477–499.

Wypler, J. (2019). Lesbian and queer sustainable farmer networks in the Midwest. *Society & Natural Resources*, *32*(8), 947–964.

7

THE 'GOOD FARMER' IN COMMUNITIES OF PRACTICE

The basic tenet of the 'good farming' concept is that farmers produce, recognise, and critique each other – and themselves – on the basis of symbols of good farming. While most farmers will seek to be recognised as 'good farmers' not all farmers achieve this status. There are a variety of reasons for this. A recent survey of farmers in Scotland, for example, found that 93% of respondents agreed with the statement "I strive to be a good farmer".[1] The few who did not agree with the statement, largely did not see themselves as commercial farmers: they identified themselves as estate managers or hobby farmers. As demonstrated in Chapter 3, symbols of 'good farming' are remarkably consistent in some respects: uniform, weed-free fields have become a standard symbol for cereal producers; freedom from weeds indicates skilled use of pesticides; uniform fields indicate skill with planting machinery and soil maintenance. For livestock producers, skill is evident in the bodies of the livestock they produce, both in terms of conformation to breed standards and overall health. For most farmers, a tidy homestead features in symbols of good farming – good farmers are those with sufficient labour available to tidy their properties and invest in appropriate housing for their animals and equipment. However, within these definitions there are considerable regional variations. Gray (1998) and others have provided theoretical explanations and empirical evidence for the consubstantial nature of farming culture – co-development of farm household, land, and the commodities they produce. In this chapter we go beyond the cultural capital formed in 'communities of place' to consider how it is developed in 'communities of practice': groups with shared practices and interests (e.g. Wenger, 1998). We do so by explicitly considering the relationship between cultural capital, knowledge, and learning, and how the formation of good farming ideals is influenced both by relations between farmers, and between farmers, industry professionals, and the general public.

Being a 'good farmer' is about the prestige given by the possession of cultural capital, which is acknowledged as valid by others. It is thus about skills and knowing: knowing how to farm well and knowing the legitimate criteria for defining what is 'well'. This knowledge is given its legitimacy by farming peers. In the literature on farming, issues of knowledge are generally understood in cognitive terms, as the ability to apply new techniques and to understand the complexity of the world (e.g. of the interaction between ecological issues and farming practices). The complexities of farmer learning and knowledge are developed in a burgeoning literature on AKIS (agricultural knowledge and innovation systems), which recognises the embeddedness of farmers in broad knowledge and learning networks, involving education, extension, and research (Dockès et al., 2011). However, knowledge in a broader sociological sense is more than skills, it is constitutive of how someone sees the world (see Carolan, 2006).

Issues of knowledge creation relate clearly to the cultural dimension of farming, but few connections have been explicitly made between the creation of new knowledge and the transformation of collective farmer identities (Forney, 2016). Bell (2004) insists on the relationship between new knowledge and new values in the emergence of individual farmer subjectivities. Within the 'good farmer' literature, the connection to Bourdieu's theoretical framework raises the question of knowledge: the socialisation processes that enable intergenerational transfer of cultural capital necessarily involve learning (Burton and Paragahawewa, 2011). Use of Bourdieu's approach also recalls the collective definition of what is recognised as knowledge. Not just any type of knowledge or skill makes a farmer a 'good farmer' from the perspective of his or her colleagues. However, little is known about how and under which conditions new knowledge and practices receive collective symbolic valuation and start to change wider cultural definitions of farming (Forney, 2016).

To date, analyses of the 'good farmer' (from the Bourdieusian perspective) have focused on deeply engrained social norms (i.e. the pattern of classical socialisation from childhood, characteristic of family farms). Resultant symbols are developed through long-term farming practice, and thus reflect the specific characteristics of the farming system (e.g. livestock farmers will have somewhat different good farming ideals than arable crop farmers). Similarities are related to production, and linked to this, the financial viability of the business (e g. different types of farmers value tidy farmsteads, healthy livestock of good conformation, and, even, weed-free crops). These symbols are visually evident (and can be critiqued by other farmers on the basis of 'roadside farming'), representing skilled role performances. Using the concept of symbolic capital, apparent 'resistance to change' amongst farmers (e.g. in relation to afforestation – Burton, 2004, or agri-environmental measures – Burton et al., 2008) is reinterpreted as resilient cultural capital – farmers gaining or retaining prestige from skilled role performance, evident in fields, livestock, and farmstead. Engagement in activities such as agri-environmental schemes is thus resisted because the activities represent a loss of cultural capital – the financial rewards of scheme participation are insufficient to outweigh the social and cultural value of maintaining long-held standards of farming practice.

The 'good farming' standard is thus explicitly identified as a shared standard between all farmers in a locale: it is located within communities of place. There are two issues here: first, technological advances have dramatically increased farmers' access to other farms and farmers across geographical space. Farmers are no longer restricted to viewing the farms in their own locales. Second, the 'good' part of the good farming concept implies that some farmers achieve this standard more than others. The 'good farmer' identity thus has a social boundary making role. To adhere to the values of the 'good farmer' means to identify as a farmer (and often to be identified as one), as a member of a farm community, and of a professional group. This collective sharing of symbols and values of reference is essential in social dynamics described in the previous chapters. At the same time, this identification as a farmer (aiming at being a 'good farmer') is inevitably constructed in opposition to other categories of people. Beyond factual and descriptive aspects (e.g. to raise animals or cultivate land, to be registered as a professional farmer), the differences between farmers and others often draw on more general moral values that allow comparisons. As an example, hard work has been identified in several contexts as a fundamental value characterising farmers' identities. By valuing a dedication to work specific to them, farmers implicitly or explicitly build a moral distinction between themselves and others who supposedly do not work as hard. The 'good farmer' serves here as a mirroring tool. Narratives produced in its name about non-farmers can be understood above all as a reaffirmation of the belonging to a community of values and practices. This belonging dimension of the 'good farmer' is perhaps more directly emphasized by theorisation of ethos and habitus which supposes an embodiment of social identities fundamental to social distinction (Bourdieu, 1984).

As seen in the previous chapters, other theoretical framings insist more on the role of the 'good farmer' reference in interaction between peers. The 'good farmer' understood as social, cultural, and/or symbolic capital (e.g. Burton, 2012; Sutherland and Burton, 2011) helps us to understand action as a way to prove oneself in front of others who share the same frame of reference. This approach includes a sense of competition, where individuals aim at the best possible performance (while avoiding bad performances), to be not only 'good enough' farmers, but 'better' farmers than neighbours or acquaintances. A farmer who has proven him/herself to be 'better' than others might become some sort of model for colleagues. A classic example would be a farmer who seems to implicitly initiate hay making in an area, with others waiting on him and trusting his decision. From this perspective, the negotiation of the criteria that constitute a 'good farmer' and their enactment is a social field of positioning, competition, or at least comparison, between individuals and/or subgroups. By extension, it helps to describe the dynamics of a structuration of the social field of farming, in its diversity.

The 'good farmer' approach was initially utilised to describe a rather homogeneous definition of performance and skills in farming. Decades of productivism led farmers' interpretations of the 'good farmer' to be formed in relation to criteria of economic success. This general framework was adapted and translated in farm communities in relation to local practices and contexts, and types of agricultural production (crops, animal husbandry, etc.). Overall, what was a successful farmer in the 20th century and how to achieve this standard was largely unchallenged by farmers. Ensuing contestation has arisen from changes in public policies, primarily in relation to ongoing market deregulation and the increasing importance of environmental schemes, but also in relation to consumer concerns regarding farming practices. The consensual and normalised 'good farming' standard disappeared with the progressive abandonment of the post–World War II productivist model. The normalisation of alternatives such as organic farming, the growing importance of administrative skills to perform in a context of bureaucratisation of agricultural policies, and increasing economic pressures in relation to free-trade agreements similarly reduced the internal consistency of 'good farmer' ideals. These factors resulted in the emergence of diverging definitions of the 'good farmer' (Sutherland and Darnhofer, 2012). Symbols of the 'good farmer' not only changed (see Chapter 3); they diversified to reflect multiple and competing ways of being a farmer in a given time and space. In essence, the definition of the 'good farmer' – from a farmer's point of view – fragmented in a context where there was no singular way to succeed in agriculture.

The work of defining – or redefining, in a changing context – the criteria that make a 'good farmer' is not limited to farmers themselves. Other actors also participate in this cultural negotiation of the good farmer. Industry professionals, providing services or goods to farmers, also develop their categorisation of who is a 'good farmer' and why. Because of social proximity to the farming community, many of those actors share symbolic understandings of the dominant definition of the good farmer. This is notable in the case of people from farming families who did not take over the family farm, but have chosen a related profession (such as cheese maker, butcher, or agro-industry salesman). Often, they master specific skills that are valued and acknowledged by farmers in relation to a role of adviser or specialists in the AKIS. In addition, these actors generally circulate widely through the farming community, and with them information about what other farmers do. Their specific positions also contribute to legitimising their perspectives. Consequently, advisers, veterinarians, salesmen, and others join farmers themselves in the constitution of specific communities of practice – groups who share a concern or passion, rather than a geographical space (Wenger, 1998) – where definitions of the 'good farmer' are discussed, enacted, and embodied, in dialogue with specific agricultural practices and the related processes of learning and knowledge creation.

In this chapter, we explore a set of examples where definitions of the 'good farmer' are contested, reformulated, diversified in specific communities of practice. This builds on the work in Chapter 3 on symbolic change but, in this case, we focus not on the symbols themselves but on the processes of social construction within farming communities. To explore this issue we first show how organic farmers contributed to reframe symbols of 'good farming' in England. Second, we consider the role of new entrants to farming in the rejuvenation of crofting culture in Scotland. Third, we look at the history of pedigree livestock farmers in Swiss dairy farming. Fourth, we consider recent research into how professionals in the agri-food sector – specifically veterinarians – understand and utilise their own standards of 'good farming' to determine what type of knowledge is legitimate and valuable. We then consider the role of the public in shaping expectations of 'good farming'. In doing so, we build a dialogue between three dimensions of the 'good farmer' concept: its embodiment in physical symbols; its embeddedness in specific locale and communities; its relation to the question of knowledge in agriculture.

Organic farming and the fragmentation of good farming identity

Sutherland and Darnhofer (2012) identified several processes that underpin the negotiation of good farming symbols. They drew on an English dataset of 48 qualitative interviews with farmers in 'hotspots' and 'coldspots' of organic farm conversion in 2007 and 2008, to argue that conversion to organic farming involves renegotiation of locale-based standards of 'good farming'. The first process is the devaluation of current symbols. For the farmers in their study, the symbolic value of high yields was devalued through their disconnection from profitable farming, resulting from increasing pressure on production costs and product prices. Similarly, the new recognition of the connection between current practices of intensive agriculture and negative outcomes, such as a disease outbreak, or the pollution of waterways clearly unsettled classical symbols by showing publicly the darker side of former agricultural 'good practices'. The devaluation of traditional symbols of good farming in their study sites led to an openness to consider other farming options, particularly reducing input use (i.e. not organic farming per se), and also experimentation with other farming opportunities (such as diversification). In doing so, farmers enter different 'fields' with associated rules by which they must learn to play (e.g. how to market to new consumers, produce new products). The devaluation of existing symbols also leads to a period of reflexivity – active consideration of what they value about farming (i.e. farmers' personal and household goals). The habitus (disposition to act) is no longer unconscious but actively scrutinised. New rules of the game thus form a crucible in which new ideals can be developed and tested; farmers may choose to play in different fields, adapt to the changing rules of their current field, or both. In this process, new communities of practice are created, offering alternative definitions of what makes a good farmer.

Utilising the same dataset, Sutherland (2013) addressed how symbols can be different and contested within locales. Organic and conventional farmers held different standards, although there were similarities. For example, both included the environment: organic farmers prioritised a varied farming landscape, with multiple crops, over the monoculture favoured by conventional producers. Skilled role performance for organic farmers was the recognition of the value of diversity and particular techniques associated with organic farming, whereas the skills of conventional producers were in pest control. Both organic and conventional farmers characterised themselves as 'progressive'. Although there was some 'if it works for him that's fine' discourse, the ideals conflicted – farmers who had not converted to organic production identified the loss of ability to treat livestock in a timely manner, or to control weeds effectively, as important barriers, some going so far as to quit farming rather than convert. In the words of 'Luke':

> I looked at organic before we gave up but now I wouldn't fancy it. I didn't like the fact that it seemed to me that you got penalised for treating a sick animal. I mean if your child had an ear infection you wouldn't hesitate to give them antibiotics if that was the right thing, but then if your cow had some kind of critical thing and you chose to use antibiotics you would get penalised through a lengthy spell where you couldn't sell the milk from that animal. And it seemed wrong to me.
>
> *(p. 437)*

For Luke, good farming included a moral obligation to the welfare of his livestock, who he likened to members of his family. It was better to leave farming than to fail to farm to his ethical standards, that is, to be a bad farmer, according to the communities of place and practice to which he belonged. In contrast, organic farmers in the study reported their formation of new communities of practice, travelling outside of the region to learn techniques from other organic farmers and to participate in organic farming demonstrations. They also identified their customers as an important frame of reference. Direct marketing to consumers enrolled the perspectives of those consumers in the farmers' assessments of their practices. For 'Gary', "producing something that somebody wants" reinforced the value of his new organic farming practices.

Sutherland (2013) argued that the symbols characterising cultural capital in contemporary farming largely reflect necessity – weed-free fields represent higher productivity and thus greater household income; healthy animals are more likely to reproduce and gain maximum value at auction sales. The higher value for organic products shifts this implicit metric – maximum value is achieved for standards of practice, rather than the quantity produced. The link between economic and cultural capital remains readily apparent, but the overt symbols of this relationship change. 'Good farmers' are successful farmers, who maintain a profitable farming enterprise, which extends to production of premium products. However, it was also clear that this connection, while direct, is not the only issue – farmers seek ideals to aspire to beyond the financial. Their discourse about farmers who are solely profit oriented is overtly negative. 'Good farming' thus lies on a continuum between necessity and luxury. Formerly commercial farmers who opted to continue farming on a recreational or hobby basis thus identified their actions as a step towards luxury – the opportunity to engage in a valued activity without need for commercial gain.

New entrants and the formation of 'good crofter' identity

The participants in Sutherland and Darnhofer's (2012) study were largely long-term farmers who had inherited their farms. In this section we consider how farming symbols are formed by newcomers to agriculture. New entrants to farming by definition are not socialised into good farming norms, nor do they gain the skills to demonstrate skilled role performance as children. They also actively choose to enter the farming profession, rather than inheriting a farm; their definitions of 'good farming' can thus be expected to reflect the privilege they perceive of being able to farm.

New entrants are particularly interesting from the perspective of understanding differential farming norms: they need to play by the same 'rules of the game' as other farmers, if they are to gain income from agricultural markets and establish positive working relationships with their neighbours, but they do not have the family background or socialised experiences of incumbent farmers. This section presents findings from a study of new entrants to crofting (a legally protected form of small-scale farming in the highlands and islands of Scotland – Shucksmith and Rønningen, 2011). Interviews were conducted with 25 newcomers to crofting (including both successors and ex novo new entrants), and nine key informants in 2013/2014. The research more broadly looked at the formation of new knowledge networks (see Sutherland et al., 2017); study participants' definitions of 'good farming' are thus embedded within broader discussions of learning processes.

The findings of the Scottish new entrants' research are consistent with Gray's (1998) concept of consubstantiality – the interconnection between the farm family, livestock, and land comprising the territory of the farm. Land quality in crofting areas is typically poor, but variable – new entrants thus clearly gauged their assessment of the quality of their own and others' production in relation to the capability of the land. Rather than absolute terms, standards of 'good crofting' identified by study participants are evaluated on the extent to which the land is used to its 'full potential'. Land abandonment – relatively common in remote crofting areas – was viewed particularly negatively.

Similar to long-term farmers, new entrants based their assessments of the potential of land on historic performance. However, as they lack lived or family experience with past productivity levels, these standards were often based on photos of the area.

MIKE: I think personally just looking at pictures of the old crofts, how crofting was around here up until maybe 40–50 years ago when you look at the well managed crofts and hay growing, people taking hay off the crofts, I can't think of many crofts around about that I would like my croft to be like that in the future.

INTERVIEWER: Is that because a lot of them aren't used?

MIKE: Yeah! Basically yeah or used to their full potential, I would say. Ours certainly isn't at the moment. but I'd like it to be like that.

Rather than accept the status quo as the 'right' or 'good' way to manage the land, study participants opted for an historic vision – visual images from 50 years ago. Instead of learning from the dubious practices of their neighbours, they sought historical sources on which to base their visual appraisal. They thus embedded their farming practices in historical approaches, but in a different form than those of their neighbours who had inherited crofts. There is nostalgia in the assessment – becoming a crofter is often about joining a traditional way of life – which links to productivist views of how crofting land should be managed. The purpose of the croft as productive is primary in the assessment.

Study findings are thus also consistent with the importance of visible symbols and road-side farming practices (visual appraisal while driving through the countryside), that is, the idea that "if you look at their croft you'll know!" Remote landscapes can make this appraisal difficult:

ALAN: Its difficult on Skye, you don't really see, you don't really see other people's places, some people's farms we'd just love to see.

TOM: We just know, we know of them and we watch their sale, their animals at sales and things, that's all, it's very much, very much second-hand information.

These respondents were describing how the mountainous terrain of Scotland's Isle of Skye made it difficult to view crofting practices from the roadside – croft housing and livestock are often located in remote areas, where there are few roads, or reasons to travel those roads. Whereas farmers in more central locations might drive past their neighbours' fields and farmsteads on a regular basis, this was not the case on Skye. Visual appraisal was thus more achievable at industry events, such as livestock auctions. Auctions in particular offer the opportunity not only to link farmers directly to their livestock, but to assess the economic value other farmers place on that livestock through the purchase price. Events like auctions also make it possible to assess the reputation of different farmers, on the basis of how their livestock are valued.

New entrants to crofting were also asked if there were established crofters who they seek to emulate. Responses were typically hybrid – no single croft or crofter was identified, but typically a combination – for example, the animal husbandry skills of one, the soft fruit production skills of another, and the diversification activities of another.

> On the technical side on the sort of farming I do talk a lot with [farming couple] at [croft name] on vegetable production, on soft fruit production, there's a very good friend and colleague in Edinbane that I talk to . . . and then there's [female crofter] who basically has taken, got into processing . . . [about] the whole processing, processing onsite.

In essence, the standards being set for new entrant crofts exceed the standards apparent in any single, existing croft. New crofters formed distinctive knowledge networks for each of their activities (Sutherland et al., 2017), establishing links to differing communities of practice. Relative 'good crofting' status was assessed in relation to similar crofts – not only in terms of land quality, but in the scale of holding and types of diversification or other activities. Crofting has historically been pluriactive – historically, crofts were allocated to the workers on estates to enable self-provision. Crofts were never intended to provide a full-time income to the resident household. Off-croft income and diversification are thus accepted norms within crofting communities. The symbols of a new 'good crofter' reflect perceived excellence in this set of endeavours.

Actively constructing a new good crofting identity involved identifying and pursuing distinction through the symbols of success for the type of croft they were developing (e.g. a good crofter as a good horticulturalist, a good gardener, a good host, or a good stockperson). At the same time, there was an evident aim for distinction – feeling that other crofts have not been successful and wanting to make a mark (i.e. to not do what everyone else is doing). This ambition is somewhat divorced from pragmatic economics, reflecting the personal preferences, innovations, and ideals of the new crofter:

INTERVIEWER: Is there anybody else out there that you kind of look at and go 'I would really like my croft to be like that some day?'

JOAN: Well it's actually a shame, there is a lot of unused crofts around, but I wouldn't say there's one that I think that's the way I want to be like theirs. I think I want to just establish my own take on the croft so my own personality, I have an idea in my head about how I want it to be so that's really what I'm working to is my own kind of ideas that I want and not really someone down the road. And certainly there are a lot of things that I want to try that I'm not seeing happening there so . . .

INTERVIEWER: So where have those ideas come from then?

JOAN: Well I've been subscribing to the *Country Small Holding* magazine for a number of years now so I get that through every month, so ideas in there, or I might be on the internet, I've got a couple of self-sufficiency books that I've bought you know just reading up and self-educating really. I've been on a few of the Scottish Agricultural College courses I'm a member of them, so just attending some courses and just getting a feel for that.

These ideals are based on a wide geographical range. It was not unusual for new entrant crofters to identify exemplar crofts from other regions of the United Kingdom or other countries (e.g. Switzerland, South Africa). Learning was frequently book and media based, in addition to practical experience of 'learning by doing'. Many of the new entrants had experience of observing or working on smallholdings from other regions, and utilised these frames of reference, rather than the actions of their neighbours. Magazines and the internet were also sources of ideas: the new entrants in this study actively constructed communities of practice which often diverged widely from communities of place.

Embedded in these symbols were standards of crofting remarkably consistent with those of farmers in other good farming studies: that land should be worked, that crofting is active (hard work, time invested), and that crofting should be profitable. Respondents were freer in their criticisms of other crofters than was the case in interviews conducted with farmers (i.e. Sutherland's other studies). This may reflect the amount of land abandonment in the study site, particularly by elderly occupiers who are no longer able to work the land, but also young people who were primarily occupying the housing without making use of the land base. While locating themselves within the heritage of crofting, new entrants see themselves as modern:

ANNE: I don't really think there's any great forward-thinking crofters in the area. . . . I think in my particular area there's only three and they're all traditional really. They keep sheep, and one of the guys is just, he's like in his late 70s or 80s, if you go to his house, his sheds it's like stepping into a time warp; its absolutely fascinating. It really is fascinating and when he dies there's nobody that I know that is a traditional old school crofter anymore.

INTERVIEWER: Uh huh. It's interesting because the new generation is so different coming through.

ANNE: Yeah! Well this guy I mean he still speaks Gaelic and he's just . . . he's fascinating! But there's no . . . in my area I don't know of anyone diversifying, sort of going forward into modern, more modern thinking.

Contemporary new entrants to crofting in the study thus expressed a critique of more traditional approaches evident in their locales, drawing their ideals of good farming from communities of practice. Several respondents indicated that none of their neighbours could be considered 'good crofters' because their land was essentially abandoned. Even in cases where land was being used, sole reliance on the traditional commodities of beef and sheep was unpopular. 'Contemporary' new entrants to crofting identified good farming as involving new commodities, consistent with their community of practice. At the same time, new entrants are seeking to ground their practices in visual images of practices from 50 years ago. Ideals are eclectic, justified in part by the pluriactive history of crofting, but also reflecting responses to contemporary 'rules of the game' for establishing financially viable smallholdings in remote areas.

Good farmers and good cows in the show ring

The 'good farmer' literature was initially developed in the context of crop farming: communities of place and practice where similar crops were produced. Therefore, the exploration of the physical and biological symbols of good farming that have been identified and analysed are mostly related to fields and their aesthetics in the production of 'tidy' landscapes. However, in farming cultures centred on livestock, such as dairy farming, the animal body has been similarly invested with symbolic values reflecting the quality of their owner and tender. Through the animal body, values and ideals are expressed and enacted, building a definition of a 'good breeder' that is shared among people involved in the same practice, but which may also be contested by individuals or groups defending other perspectives on farming and animals. As with 'roadside farming', and as mentioned in the previous section with the example of livestock auctions in Scotland, the visibility and exhibition of symbols here plays a central role in the building of farmers' reputations and identities as 'good farmers'.

Historically, there were many symbols associated with the bodies of animals that indicated good husbandry or good breeding. In terms of breeding, farmers looked for qualities of the animal that were believed (rightly or, as was often the case, wrongly) to indicate secondary qualities of the animal associated with either the ease of management or the quality/quantity of produce. For example at the end of the 18th century in the United Kingdom, the form of the animal could indicate the disposition to fatten, a small head was a sign of easy birthing, while having small bones and a small proportion of offal were considered indications that the animal would provide a large quantity of meat (Sinclair, 1802). Slightly less credibly, the silkiness and elasticity of the skin was believed to indicate a tendency for the animal to "take on meat" as it could be "more easily stretched out to receive any extraordinary quantity of flesh, than a thick or tough one" (Dickson, 1807; Sinclair, 1802, p. 406), while farmers were advised to avoid white animals by one of the key agricultural improvers of the time as, he believed, white was "a sign of degeneracy in all the animals of the creation" (Young, 1791). Other symbols indicted good or bad husbandry on the part of the farmer. For example, smooth skin (Clarke, 1781) and a lack of ticks (Collins, 1799) were indications the animal had been well cared for. On the other hand, "a belly more capacious than common" on weaning calves symbolised that the animals had been underfed (Batchelor, 1808).

Today breeding societies, shows, and pedigree livestock sales offer places and times where symbols of what make a good and healthy farm animal are staged and made explicit. As we will see however, the participation in such events has progressively become a specialised practice, limited to a specific community of breeders, at least in Switzerland. As travel has become more efficient, and technologies progressed (e.g. enabling tele-purchasing), shows and sales have shifted from being central to communities of place to spaces where communities of practice can gather and exhibit their good farming credentials. Show animals in particular become physical embodiments of good farming prowess. Farmers make critical distinctions between 'working' livestock (those which perform well in commercial herds) and 'show' livestock (which perform well in show rings and at auction). The latter reflect the 'taste of luxury': characteristics typically reflect but do not identically mirror the symbols of efficient commercial production (e.g. show livestock may be high maintenance in that they require additional feed or grooming to achieve success).

For this section, we draw on data collected in long term ethnographic fieldwork (including participant observation, semi-directive and informal interviews, and historical and contemporary document analyses) in the French-speaking part of Switzerland, mostly between 2002 and 2010, in the context of a rapidly evolving dairy industry. The history of dairy farming offers a paradigmatic example of the evolution of the symbols of the good cow. During the 20th century, the development at regional, national, and international scales of technological and institutional tools for improving the quality of animals (such as herd books, standardised metrics of evaluation, shows and competitions), produced very strong symbols, shared widely in the 'modern' agriculture of the global North. This does not mean however that the symbols and narratives of the good cow are all similar independently of context. Holloway (2005, p. 883) has shown how diverse technical criteria of evaluation represent diverse types of "constitution of bodies embedded in specific nature – society", when comparing aesthetic and technical knowledge in cattle breeding in the United Kingdom.

Dairy cattle breeding has been an important part of farmers' pride and fame for many decades in Switzerland. Local breeding societies were created at the end of the 19th century or just after. They have been crucial actors in the differentiation and consolidation of distinct breeds and in promoting emulation between breeders. They created new spaces and practices of institutionalised competition – such as yearly local and regional shows, where farmers could stage and perform their skills as breeders. They also produced new metrics for assessing animal bodies according to new standards of productivity and representations of the ideal cow. On this basis, the good cow embodies the good farmer's identity and secures it over time, first through the life of the individual animal, but also and primarily through the genes she will pass on to the next generation. Being a good breeder is then an identity that forms over time, and often through generations of farmers.

In recent times, detailed representations of the ideal cow have evolved together with farming practices, and agricultural economics and politics. In this sense, the animal body is not only the physical symbol of the quality of the individual farmer, it also expresses an agricultural model. At the end of the 20th century, this model was based on productivist values. As with tidy landscapes, or well-maintained and productive crofts, these breeding practices and values are translated into aesthetics. A good cow is also a 'beautiful' cow. Or better said, what makes a cow beautiful is the visible expression of her productive capacity. In Switzerland those criteria have been at the same time unified and specified for each breed, in relation to where they are farmed. The Swiss Simmental, a historical local breed, represents a diversified mountain agriculture, combining both meat and milk production qualities, associated with a robustness adapted to the rougher conditions of mountainous regions, while other breeds in other regions specialised univocally in dairy. The Friesian Holstein breed, a worldwide ambassador of high productivity dairy cattle, embodies a more intensive and specialised style of dairy farming. In some regions of Switzerland, this international breed has become a substitute for local breeds which have disappeared.

Breeding societies have been active players in those processes. They continue to exert influence through herd books where all the cows of every farmer member are registered, and their performance recorded: production, quality of the milk, fertility, and genealogy. In addition, experts visit farms assessing visually and measuring the morphology of the animals, following a standardised process of 'linear description'. Recently, gene technology has strengthened the potential for selection (Holloway and Morris, 2012). Genomic selection makes possible the genetic identification of specific characteristic and assessing the genetic fit with the 'breed type' and still increases the capacity of breeders to shape the animal body according to their objectives and representations of the good cow.

In parallel to these increasingly quantified and precise technologies of assessment, farmers can present their best cows in competitions. Interestingly, the ranking at those competitions traditionally depends on the decision of a single judge, who decides purely on the basis of visual evidence. Bodily conformation to the specificity of the breed is here the only official factor of selection and assessment. For dairy cows, good teats and mammary system are crucial elements. Size is not the main element, but good blood irrigation, practicality for milking, and a strong ligament for longevity are essential. Dairy cattle competitions therefore have specific ranking and prices related to teats. Other factors are also considered, such as the quality of the legs and a strong and straight spine for a good health, a wide rump for easy calving, etc. That this assessment is based exclusively on visual factors is somewhat surprising, notably because of the discordance between increasingly technologically mediated assessments used otherwise in the breeding societies (Holloway and Morris, 2012). Show-ring aesthetics draw on embodied knowledge about what makes a good productive cow. All of the visual and aesthetic criteria have a rational explanation in terms of animal health and productivity, inspired by a productivist ideal of the animal. This said, differences between animals are subtle and only a trained eye can make appropriate distinctions. This is maybe where those competitions participate more fundamentally in the constitution of a collective identity: in the creation of shared skills in this 'skilled vision' of animal bodies (Grasseni, 2009), not only as breeder, but also as judges, and in the expression of a culture of animal farming.

Competitions are organised at multiple levels: local, regional (the Canton), and national, in accordance to the federal organisation of the societies (local societies are members of regional societies, themselves regrouped within a national federation). In the recent past (i.e. the end of 20th century), every dairy farmer would participate in the local, and most in the regional competitions. The gatherings of those communities of place are unique opportunities to compare and situate one's performance as a breeder. They are also a special time for socialising. Farmers also participate for recreational purposes, identifying shows as a 'hobby', and for their personal enjoyment of seeing 'beautiful cows'. However, such competitions are also unique opportunities for marketing and promotional activities: showing the quality of your cattle can increase their sale value. Performance in breeding is supposed to translate into income through animal sales, other farmers wanting to invest in good cattle and being ready to pay a good price for a genetically superior cow. Competitions are also show rooms for famous breeders and an arena to prove yourself as a breeder. This fame will, in addition to the recognition and pride, result in income, even if the prizes won in the competitions are, at least in the Swiss context, symbolic (e.g. a decorated cow bell).

However, in recent years there are indications that what was a largely shared culture is becoming more and more specialised and reduced to a smaller number of participants, at least at the level of active participation. The decreasing number of farmers make local competitions less significant and many local competitions have disappeared. Regional competitions remain accessible, but national competitions are often opened to international participants and the competition has reached levels that are unachievable for 'normal' good breeders. Indeed, some farmers have specialised in breeding and competition, forming a new, very competitive community of practice. What was an accessory activity to dairy farming has become a new business model where farm income might well come more from the selling of the cow's genetics (in the form of animals, sperm, or embryo) than her milk. At the same time, commercial dairy farming in Switzerland has moved away from a classical productivist model where production would equal profitability. The double move of progressive market deregulation and greening of agricultural policies has dramatically changed the rules of the game and farming practices. Having to face increasing market pressures and production costs, standard dairy farmers have neither the time, the capacity, nor the economic interest to fight in this competition anymore. The most 'beautiful' cow according to show ring criteria is no longer necessarily the best cow, as profitability, the new master word in dairy farming, has become a complex matter where productivity is just one factor. In the making of the good cow in the Swiss dairy farming communities, a new rupture has appeared between the 'ideal cow' related to the practice of fewer and fewer accessible shows and competitions, and the 'pragmatic cow' on farms, made of compromises and trade-offs between production costs, the farm business model, and a continuous struggle for profitability. While the aesthetics of the good cow initially developed as a symbol of good farming and economic success, there is a need now, in the farmers' words, to balance 'beauty' and 'economics'. In this process, the symbolic value of the animal's body, as incarnation of a productivist ideal, has lost a substantial part of its strength.

Good farming for agricultural industry professionals

The previous examples have focused on the diversity of the definitions of the 'good farmer' in relation to different farming communities of practice. However, as pointed out in the introduction to this chapter, some non-farming actors might also play a specific role in this process because of their liminal position, made of frequent professional interaction with farmers, but a clear non-farmer identity. Through their expression of 'expert' knowledge, industry professionals can reinforce or challenge farmers' conceptions of what it is to be 'good'.

Recent research related to biosecurity issues offers useful examples of interactions between veterinary services and farmers around the definition of good farming. Some research focuses on farmers' discourses and practices defining 'good farmer' identities in disease management, as for instance Naylor et al. (2018) who identify three good farming ideals in relation to biosecurity: livestock husbandry, being a good neighbour, and the good public-facing farmer. Research into the veterinary understanding of the 'good farmer' points more clearly to the interaction between vets and farmers regarding the definition of good farming. As an example, in her unpublished PhD thesis, Middelveld (2019) found that many Scottish farmers prefer to pro-actively treat for sheep scab, the disease at the centre of the research, when symptoms are first identified. Farmers had learned how to effectively treat their sheep and proceeded to do so without veterinary assistance. Farmers who responded in this way identified the skills involved in recognising sheep scab (i.e. skilled role performance) and the importance to them of preventing sheep suffering (i.e. through early treatment, similar to the perspective of Luke, earlier in this chapter). Thus, in this community of practice, early treatment was generally understood to be good farming. Interestingly, veterinarians also bought into this line of reasoning – selling the required medications over the counter without a vet visit to the farm. Although sheep scab is a notifiable disease, these vets do not notify the authorities, as no sheep scab has been formally identified. The rationale provided is that the farmers are clearly managing the situation successfully. In other words, for veterinarians too, a 'good farmer' is capable of identifying sheep scab and treating it without veterinary assistance; they would only see a need to intervene if the scab went untreated, even if the current regulation would require them to report the disease immediately (i.e. before any treatment). In this case, vets and farmers apparently share similar definition of the 'good farmer' within the community of practice they constitute.

Middelveld's (2019) findings on sheep husbandry in Scotland contrast with those of Shortall et al. (2018) on another community of practice: dairy production in England. Shortall et al. (2018) found two conflicting 'good farmer' identities in relation to biosecurity, one based on farmers' husbandry skills and the other lodged in vets' scientific expertise. In her study, vets identified as good farmers those who invested substantively in veterinary services. These were also large-scale, commercial farmers who could afford these services. Vets were engaged to lead biosecurity and herd health programs, which included both treatment and prevention. In contrast, farmers who were not heavily reliant on veterinary expertise for biosecurity management prided themselves on their skilled role performance. From their perspective, being a 'good farmer' included the ability to diagnose disease through visual appraisal. While some vets appreciated the skilled role performance on which farmers prided themselves, others pointed out the limitations when dealing with diseases which were in carrier state (i.e. transmissible but not visibly symptomatic). However, detection may be dependent on the abilities and experience of the farmer. In a separate study, Burton et al. (2012) observed that farmer's close and regular contact with dairy cows enabled them to observe minute and subtle deviations from normal well-being – explained as "you just sense something" or "you can just tell" – before symptoms of the condition fully emerge and are diagnosable. The conflict reflects valuation of different types of knowledge: tacit knowledge, gained by farmers through direct experience of caring for livestock, and the explicit, codified 'professional' knowledge of veterinarians learned through formal training.

The conflicting definitions of good farming also reflect a shift in veterinary models of practice from 'testing and treating' to 'predicting and preventing' (Sibly, 2010 in Shortall et al., 2018). A change in veterinary standards of what it is to be a good vet was impacting on the way vets were characterising 'good farmers'. It also reflects intensification within the industry and the changing farm practices that were associated (larger herds, with stock sourced from broader geographical distances) and the heightened biosecurity threats that result.

Both Middelveld (2019) and Shortall et al. (2018) raise the issue of stigma associated with livestock disease. Farmers are reluctant to discuss the incidence of disease, as this implies a failure to meet the good farming standard that includes the notion that 'only bad farmers get diseases' (Heffernan et al., 2008). Consequently, good farmers are those without disease, or those who 'prevent' it. In Middelveld's thesis, she found that farmers treating sheep scab in the early stages identified the treatment as 'preventative' – although the symptoms of sheep scab were clearly present, and used to justify treatment, farmers did not consider the sheep to have the mites. Incidence of sheep scab was thus non-existent to both the farmers and the veterinarians from whom they bought their supplies. In Shortall's research, farmers were reluctant to discuss biosecurity, or to address specific issues, in part because of the resultant loss of cultural capital. Indeed, purchase of livestock at auction, as argued in the previous sections, is a visible act of skilled role performance: the farmer demonstrates the ability to select quality animals and the economic ability to pay high prices. Biosecurity precautions which limit the introduction of live animals into the herd (relying instead on artificial insemination) thus represent losses of social and cultural capitals.

The good cow in public debates

The role of non-farmers in shaping the 'good farmer' is not limited to agricultural industry professionals. Consumers of various aspects of farming – such as the organic produce consumers described earlier – also play a role in communities of practice. In the Swiss context, animal bodies, and more specifically cattle horns have recently been at the centre of public debates. Swiss citizens voted in 2018 on a popular initiative[2] which proposed to direct specific payments to farmers who keep the horns on their livestock (cows and goats), through the agricultural policy system. This section presents an analysis of the debate that developed between different communities of interest, which includes farmers' community of practices and other actors of the Swiss society (animal rights activists, politicians, and diverse commentators). It builds on long-term ethnography among the Swiss dairy farmers and industry (as described previously, in the section on the good cow) and draws more specifically on a non-exhaustive corpus documenting the debates in the Swiss media in 2017–2018 (radio and television programmes, newspapers, websites).

Interestingly the policy initiative was led by an individual farmer and opposed by most of the farming organisations. This situation reflects the diversity of farming practices in the Swiss context. While a few breeds are generally kept with horns, notably in the mountain areas, most dairy cows (90%) – and generally the most productive among them – are dehorned as calves. Animal welfare regulations have become stricter in order to limit animal suffering and farmers have to take a course in order to be authorised to dehorn their calves themselves. Alternatively, dehorning can be done by veterinarians. The current practice accepts implicitly that horns are an inert and decorative attribute that can be suppressed without consequence, and that it is right to do so to avoid the risk of injuries for animals and humans. As a farmer, also a member of the National Parliament, said in a radio interview:

> Even if the cows are docile, they can hurt us by moving their head, wanting to give a cuddle, with their horns, unwillingly. Also, don't forget that they hurt one another. The cow always wants to show who's the queen of the herd. . . . Finally, with this system [dehorning], they are as lively, but without hurting one each other . . . with as much pleasure of living in their free stall, with or without horns. I personally don't see any difference in behaviours.[3]

However, the idea that the horns are useful part of the animal body is defended by animal welfare organisations and minority groups of farmers. In the same radio programme, a biodynamic female farmer contests the idea horns make no difference. In her argument, she defends another agricultural model, "with herds that are not too big and cows who know us".[4]

The debate thus juxtaposed different agricultural practices related to the different farming communities. However, for the president of Prométerre, a regional farmers' organisation,[5] as for many farmers, this controversy about horns was first of all an illustration of emotional opinions of citizen-consumers being more and more disconnected from the practical realities of farming and the pragmatic constraints of agricultural production. Interestingly, the supporters of the initiative also criticised on their website the incoherence between the images of the Swiss agriculture and its reality:

> Both the postcard Switzerland and the idyllic advertising full with horned animals pretend to us [*sic*]. In Switzerland there are only about 10% horn-bearing cows left.[6]

However, the promoters of the initiative also refer to very concrete aspects to justify their claim: keeping horned cows requires more space per animal in the stable and this results in higher costs that their initiative want to compensate for through state direct payments. Money was also central in some arguments of opponents. For instance, the head of the largest farmers' union (the Swiss Farmer Union) warned that if citizens voted in these new subsidies for horns, they would have to be paid from additional public money and not taken from the current agricultural budget, which would mean from the pocket of other farmers. However, the union did not give any voting recommendation, a decision that reflected the divergence within the farming communities in relation to opposing practices and interests.

A more symbolic reading of this debate, however, can easily point to the importance of the 'naturalness' of the cow in the debate on agriculture and farmed animals. On the one hand, horned cows are associated with representations of an extensive, mountain, diversified pasture-based animal farming systems. This minority agricultural practice and the horned cows offer an iconic image of traditional alpine agriculture, strongly attached to the Swiss national identity, as well documented by research, notably in the context of the 700th anniversary of the country in 1991 (Berthoud et al., 1991). It also represents a type of agriculture that can be seen as particularly respectful of nature and animals. Therefore, advertising and marketing campaigns picturing cows in the Swiss media, for a non-farmer, urban dweller audience, typically show horned cows. A perfect example is the iconic and very popular cow Lovely, who appears in most of the marketing campaigns led by Swissmilk, the national federation of dairy farmers. Lovely proudly wears her beautiful horns while promoting the goodness of milk, but she is – obviously to any farmer – from the Holstein breed, a very productive breed that is systematically dehorned in most of the commercial farming world. From a farmers' perspective, Lovely is a marketing fake. However, in this kind of marketing imaginary, as well as in the "postcard Switzerland" mentioned in the initiative's website, horns act as a double symbol, both of traditional agriculture and national identity, and, with a more modern tone, of animal welfare and sustainable agriculture.

On the other side of the value chain, dehorned cattle represent the reality of the huge majority of intensive and semi-intensive dairy farms in Switzerland. As shown earlier in this chapter, the body of the cow in dairy farming has developed a symbolic value based on its productive function. The good cow's body is disciplined, tamed, and productive. From this perspective, horns have few positive values. Instead, they are represented as dangerous to humans and other cows. Horns symbolise the wild part of the animal, its capacity to harm and therefore to resist human control. In this sense, horns oppose the ideal of a controlled nature that has been identified as central in farmers' representation of tidiness in landscape (Burton, 2012), and that has been clearly illustrated for Switzerland by the work of Miéville-Ott (Droz and Miéville-Ott, 2001; Miéville-Ott, 2003): nature as a force to control in the farm environment. As weeds in fields or pine regrowth in a mountain pasture, horns are symbolically out of place in the productive body of the dairy cow.[7]

The initiative for subsidising cows' horns was rejected by 54.7 % of the voting population on the 25th of November 2018. However, the public debate illustrates how specific definitions of the 'good farmer' can be discussed and contested in broader fora, pressing farmers and their representative to explain and justify what they do on their farms. Doing so they make more explicit the symbolic dimension of farming practices. This vote might not have influenced directly how Swiss dairy farmers form an ideal of the cow's body as an embodied symbol of a good farmer, but it has nevertheless contributed to shape the wider social and cultural frame in which such farming cultures have to express themselves, and not only on the specific question of horns and animal welfare in farming, but also on the role of agricultural policies and regulations in the shaping of the Swiss agriculture.

Good farming, learning, and the identification of 'bad farmers'

The divergent 'good farming standards' presented in this chapter demonstrate the negotiated and situated nature of cultural capital production. New knowledge and skills develop in dialogue with the particular needs and challenges met in a specific community of practice. With them, new symbolic expressions of skills and performance take shape, while older symbols slowly become obsolete. Although symbols of good farming inevitably reflect the land base of the farm to some degree, cultural capital can be – and is arguably increasingly being – disconnected from specific farms and influenced by broader sectoral trends and networks of farm household members. Farmers are not limited to visual observation of their neighbours' fields or sales at local auction marts, which are arguably decreasing in importance as sites of cultural production. Farmers have access to images and direct observation of farms and auctions in other countries and on-line. They also have access to scientific knowledge, through the expertise of agricultural sector professionals, and their own formal education. Similarly, non-farming actors engage in the discussion of 'good farming', directly through their involvement in the agricultural sector, or indirectly in public debates. Despite their role as the physical embodiment of cultural values, symbols are always contested, discussed, challenged, perhaps more intensely at present, reflecting the changing structure of agriculture, brought about in part by low and fluctuating economic returns from farming.

The chapter thus raises the question of how knowledge and learning feature in good farming constructs. To date, analyses of good farming have steered clear of the extensive literature on learning, focusing instead on socialisation processes. Social science has a variety of contested concepts for understanding knowledge and learning, but distinctions between tacit (implicit) and codified (explicit) knowledge (traced back to Polanyi, 1958) are generally recognised. In essence, implicit knowledge or 'know-how' is acquired through practice and experience – the skilled role performance characteristic of 'good farmers' – and is not necessarily related to cognitive learning (Curry and Kirwan, 2014 – also see Chapter 3). Farmers may be able to perform a skill (e.g. identifying a sick animal) without being able to explain how they do it, much like it would be challenging for most people to explain how to ride a bicycle. Explicit or codified knowledge is more 'scientific' or factual – easily reported and documented. Cultural capital, and the ability to appreciate it, includes both forms. Tacit knowledge is essential as an embodiment of specific skills and competences that are often learned through practice and inherited unconsciously in the collaboration with parents or colleagues. However, the importance of scientific knowledge is also evident in agricultural knowledge and innovation systems (AKIS), where formal education, extension services, and specialists of different types play key roles. Moreover, competing expertise, as explicit knowledge, emanating from non-agricultural actors becomes important when an issue is publicly debated in diverse fora. Formal and codified knowledge impacts therefore on the progressive changes and re-negotiation of farming symbols – the data on dairy cattle productivity, contemporary thinking on biosecurity management, the prices for organic products and the role of horns. The shift in the 'rules of the game' from those of husbandry to business management suggests a shift in the type of skills and competences that are needed to succeed in farming, creating new types of cultural capital, which threatens existing cultural capital, engendering an apparent resistance by long-term farmers and those unwilling – or financially unable – to professionalise at the speed of farming technology development.

The changing nature of symbols thus raises the issue of 'bad farmers' – how farmers who do not achieve the status of good farmers view themselves, and their neighbours. This definition is implicit rather than explicit in the good farming literature. Evidence of 'bad farming' (e.g. dead stock left in fields) is visibly apparent to neighbours and subject to sanction. In Sutherland and Darnhofer's (2012) example of the development of an organic community in England, the question of who is a 'bad farmer' highlights divergences and potential conflict and competition between specific communities of practice: 'good practice' in treating sick livestock is contested. In relation to the development of new entrant crofter identity, newcomers actively reject the typical approaches to crofting in their locales, drawing instead on both historic images and examples from other regions and countries to formulate individualised, hybrid good crofting standards. For Swiss dairy farmers, the tensions between the ideal cow and the practical pressures on profitability open the debate on who is actually the good breeder: the one with the most beautiful cow, or the one who best balances the symbolic need for beauty with pragmatic herd management. In the research on veterinarian perspectives on good farming, farmer understanding of bad farming – leaving livestock untreated while diagnosis is officially verified – is supported by vets in Middelveld's (2019) study but challenged in Shortall et al. (2018). In the latter, 'bad farming' is understood as non-conformity to the changing standards of veterinary practice. Finally, the public debate and the intervention of non-farmers in the debate about what is a 'good farmer' in relation to keeping or removing the horns of the cows introduces new arguments and capitals in the dialog between two farming communities.

The tug of war between farmers invested in different forms of cultural capital and knowledge demonstrates the influence of changing profitability on farming culture. Success in the show ring, sales of organic produce, and establishment of a new farm are all marks of distinction linked to financial profitability. The fragmented nature of contemporary good farming ideals reflects the challenges of achieving economic success – efficient production and strong husbandry skills are no longer enough. 'Bad farming' practices may directly and indirectly reflect the financial challenges of the farming sector – economies of scale leaving little time to develop or employ husbandry skills, mis-informed practices undertaken in the name of 'organics' to achieve premium prices. Sanctions against bad farmers are not limited to reputational damage; Sutherland and Burton (2011) found that farmers in their study would not share expensive equipment with neighbouring farmers who are not deemed to be 'good farmers'. Farmers who did not appear to take care of their own farms well could not be trusted (in Bourdieusian terms, a loss of cultural capital directly leading to a loss of social capital). Communities of place thus become self-reinforcing, as social capital can reinforce – or not – the cultural capital valued by local community members. In contrast, communities of practice offer greater flexibility for the development and promotion of new good farming ideals.

Notes

1 Unpublished finding from the 2013 CAP Intentions Survey conducted by the James Hutton Institute and Scottish Agricultural College, Scotland.
2 'Popular initiatives' are part of the Swiss democratic system. With them a group of actors (citizens, associations, groups . . .) can propose a modification of the law or a new law. First, they have to collect 100,000 signatures of Swiss citizens supporting the initiative. If they succeed, there is a national vote. If the vote is positive, the government has to include the new law and integrate it in its policies.
3 Original in French: "Même si les vaches sont dociles, elles peuvent nous blesser en tournant la tête, en voulant nous faire un petit câlin, avec sa corne, sans le faire exprès. Et puis, sans oublier les blessures qu'elles se font entre elles. La vache veut toujours montrer laquelle est la reine du troupeau . . . Finalement, avec ce système, elles sont tout aussi vives, mais sans se blesser, avec autant de plaisir à vivre dans leur stabulation, avec ou sans cornes. Je ne vois pas de différence de comportement, personnellement." Radio interview RTS 1, La Matinale, 03.10.2018 [www.rts.ch/play/radio/la-matinale/audio/le-monde-agricole-partage-devant-linitiative-pour-les-vaches-a-cornes?id=9889306]
4 Original in French: "Nous, on est de l'idée d'avoir des troupeaux pas trop grands et des vaches qui nous connaissent", Radio interview RTS 1, La Matinale, 03.10.2018 [www.rts.ch/play/radio/la-matinale/audio/le-monde-agricole-partage-devant-linitiative-pour-les-vaches-a-cornes?id=9889306]
5 RTS 1, 25.11.2018, special news programme on the result of the votes
6 English version of the website: https://hornkuh.ch/en/horn-cow-initiative/
7 This has not always have been the case. Horn was historically an important material in manufacturing where it could be used as a rudimentary plastic to make objects such as snuff boxes, buttons, combs, knife handles, and powder horns. Thus, as was the case with other agricultural products observed in Chapter 3, the value of a cow was, at some point in history, not solely based on food production but also on its ability to provide other materials of value.

References

Batchelor, T. (1808). *General view of the agriculture of the county of Bedford*. London: McMillan.

Bell, M. (2004). *Farming for us all: Practical agriculture and the cultivation of sustainability*. University Park: Penn State press.

Berthoud, G., Crettaz, B., and Preiswerk, Y. (1991). *Vache d'utopie*. Genève: Éditions Slatkine.

Bourdieu, P. (1984). *Distinction: A social critique of the judgment of taste*. Cambridge: Harvard University Press.

Burton, R. J. F. (2004). Seeing through the 'good farmer's' eyes: Towards developing an understanding of the social symbolic value of 'productivist' behaviour. *Sociologia Ruralis, 44*(2), 195–215.

Burton, R. J. F. (2012). Understanding farmers' aesthetic preference for tidy agricultural landscapes: A Bourdieusian perspective. *Landscape Research, 37*(1), 51–71.

Burton, R. J. F., Kuczera, C., and Schwarz, G. (2008). Exploring farmers' cultural resistance to voluntary agri-environmental schemes. *Sociologia Ruralis, 48*(1), 16–37.

Burton, R. J. F., and Paragahawewa, U. H. (2011). Creating culturally sustainable agri-environmental schemes. *Journal of Rural Studies, 27*(1), 95–104.

Burton, R. J. F., Peoples, S., and Cooper, M. H. (2012). Building 'cowshed cultures': A cultural perspective on the promotion of stockmanship and animal welfare on dairy farms. *Journal of Rural Studies, 28*(2), 174–187.

Carolan, M. S. (2006). Do you see what I see? Examining the epistemic barriers to sustainable agriculture. *Rural Sociology, 71*(2), 232–260.

Clarke, C. (1781). *The true theory and practice of husbandry*. London: G. Robinson and T. Slack.

Collins, J. (1799). Some further practical remarks on the nature of sheep and wool; and the disorders of sheep, etc. *Letters and Papers on Agriculture, Planting, Etc. addressed to the Bath and West of England Society, 9*, 113–125.

Curry, N., and Kirwan, J. (2014). The role of tacit knowledge in developing networks for sustainable agriculture. *Sociologia Ruralis, 54*(3), 341–361

Dickson, R. W. (1807). *Practical agriculture or, a complete system of modern husbandry*. Vol. 2. New edition. London: Richard Phillips.

Dockès, A.-C., Tisenkopfs, T., and Bock, B. (2011). *Agricultural knowledge and innovation systems in transition – a reflection paper*. Retrieved from Standing Committee on Agricultural Research (SCAR) Collaborative Working Group AKIS. Brussels.

Droz, Y., and Miéville-Ott, V. (2001). *On achève bien les paysans. Reconstruire une identité paysanne dans un monde incertain*. Chêne-Bourg and Genève: Georg.

Forney, J. (2016). Blind spots in agri-environmental governance: Some reflections and suggestions from Switzerland. *Review of Agricultural, Food and Environmental Studies, 97*, 1–13.

Grasseni, C. (2009). Good looking: Learning to be a cattle breeder. In c. Grasseni (Ed.), *Skilled visions. Between apprenticeship and standards* (pp. 47–66). New York: Berghahn Books.

Gray, J. (1998). Family farms in the Scottish borders: A practical definition by hill sheep farmers. *Journal of Rural Studies, 14*(3), 341–356.

Heffernan, C., L. Nielsen, K. Thomson and Gunn, G. (2008). An exploration of the drivers to biosecurity collective action among a sample of UK cattle and sheep farmers. *Preventive Veterinary Medicine, 87*(3–4), 358–372.

Holloway, L. (2005). Aesthetics, genetics, and evaluating animal bodies: Locating and displacing cattle on show and in figures. *Environment and Planning D: Society and Space, 23*, 883–902.

Holloway, L., and Morris, C. (2012). Contesting genetic knowledge-practices in livestock breeding: Biopower, biosocial collectivities, and heterogeneous resistances. *Environment and Planning D: Society and Space, 30*(1), 60–77.

Middelveld, S. (2019). *Sheep scab in Scotland: An exploration of multiple disease situations*. Unpublished PhD Thesis. Department of Sociology, University of Aberdeen, UK.

Miéville-Ott, V. (2003). Multifonctionnalité et identité paysanne. Une étude auprès des agriculteurs de l'Arc jurassien. *Économie et sociologie rurale, 1*(3), 131–144.

Naylor, R., Hamilton-Webb, A., Little, R., and Maye, D. (2018). The 'good farmer': Farmer identities and the control of exotic livestock disease in England. *Sociologia Ruralis, 58*(1), 3–19.

Polanyi, M. (1958). *Personal knowledge. Towards a post-critical philosophy*. Abingdon, UK and New York: Routledge and Kegan Paul.

Shortall, O., Sutherland, L. A., Ruston, A., and Kaler, J. (2018). True cowmen and commercial farmers: Exploring vets' and dairy farmers' contrasting views of 'Good Farming' in relation to biosecurity. *Sociologia Ruralis, 58*, 583–603.

Shucksmith, M., and Ronningen, K. (2011). The uplands after neoliberalism? – The role of the small farm in rural sustainability. *Journal of Rural Studies, 27*(3), 275–287.

Sinclair, J. (1802). Hints regarding cattle. *Annals of Agriculture and Other Useful Arts, 38*, 389–409.

Sutherland, L.-A. (2013). Can organic farmers be 'good farmers'? Adding the 'taste of necessity' to the conventionalization debate. *Agriculture and Human Values, 30,* 429–441.

Sutherland, L.-A., and Burton, R. J. F. (2011). Good farmers, good neighbours? The role of cultural capital in social capital development in a Scottish farming community. *Sociologia Ruralis, 51*(3), 238–255.

Sutherland, L.-A., and Darnhofer, I. (2012). Of organic farmers and 'good farmers': Changing habitus in rural England. *Journal of Rural Studies, 28*(3), 232–240.

Sutherland, L.-A., Madureira, L., Dirimanova, V., Bogusz, M., Kania, J., Vinohradnik, K., . . . Knierim, A. (2017). New knowledge networks of small-scale farmers in Europe's periphery. *Land Use Policy, 63*, 428–439.

Wenger, E. (1998). *Communities of practice: Learning, meaning, and identity*. Cambridge: Cambridge University Press.

Young, A. (1791). A few notes taken in Sussex. *Annals of Agriculture and Other Useful Arts, 15*, 427–434.

8

FUTURE CHALLENGES FOR THE 'GOOD FARMER'

With this book we have brought together the literature that developed in relation to uses of the 'good farmer' concept to describe how farmers 'see' themselves and other farmers, notably after the seminal paper by Burton (2004), but not limited to his work. We have delved into the historic uses of the 'good farmer' term, analysing the early development of symbols and good farmer identities (Chapters 2 and 3). Good farming identity and judgements are clearly co-constructed with the land on which farmers work, the machinery and tools available to them, the commodities they produce, and the families and communities with whom they work. The good in farming is ambiguously related to ideas both of performance and moral goodness (Chapter 5). Because defining who is a good farmer is inevitably related to positionality, the 'good farmer' can also serve as a lens to scrutinise power relations in farming communities and highlight those whose positions dominate the social norms. In Chapter 6, we clearly see a dominance of masculinity and heterosexuality in the representation of who can be a farmer, a dominance which may in part be reinforced by the 'good farmer' literature. Clearly there is no universal definition of a good farmer. As shown in Chapter 7 definitions are multiple, contextual, and changing. They evolve in a constant dialogue within communities of practice and in interaction with wider society.

In this concluding chapter, we first return to the relationship between the social and cultural dimensions of agriculture and processes of change in agricultural and food systems. We offer an attempt to reconcile change and farmers' resistance by advocating an approach that would consider definitions of the 'good farmer' not as a barrier, as is often the case, but rather as a basis for dialogue and co-construction of futures in food and farming. As definitions of the good farmer are always challenged by changes in society, be they technical, political, or economic, we then offer an initial list of future challenges that farming populations will have to deal with in the coming years. Those processes of transformation will deeply impact what it means to be a farmer, and how to be a good one. Finally, we conclude with some suggestions about future developments of the 'good farmer' conceptualisation and potential methodological approaches.

Farmers facing the transformation of future agri-food systems

Resisting disappearance

For much of the 20th century, substantive rural policy and academic research revolved around what came to be known as the Agrarian Question: the survival and continued prevalence of family farming. Political economy analyses of the transition from feudal to capitalist agricultural production led to the predicted demise of family-based production under market capitalism (Newby, 1987). The presumption was a modernist one: of course science, technology, and industrialisation would replace those people responsible for food and primary commodity production. Both Marxist and Weberian theories supported classical economists' contention that family farms would disappear in the face of economies of scale. Although this clearly did not occur in the 20th century, in the 21st century the issue of the future forms of farming, and the role of farming families in these transition processes, has reemerged.

The importance of the Agrarian Question for this book is the peculiar nature of the agricultural sector within late capitalism, reflecting specific land and household-held capital relations. Substantive debate on the issue in the 1970s and 1980s revived Chayanovian (1927) contentions from the early 20th century. Chayanov argued for the inherent resilience of family-based production: families tend to be more flexible than employees in their labour allocations, and are better able to navigate the low and fluctuating returns of agricultural commodity production. Kautsky (1988), another favoured source, argued that land was so intrinsically valuable to peasants that they would reduce production and sell their own labour in order to retain control over their land. Seminal work by Harriet Friedmann (1978, 1980) updated these debates, arguing that there is no single 'natural' capitalist form of agricultural production, simply forms which are viable under particular conditions. Friedman pointed specifically to the joint nature of family-held farms – that they are at once both a household influenced by broader socio-economic conditions and a unit of production. Although aspects of agricultural production inevitably enter markets through modernisation, risks to profitability associated with weather patterns and volatile commodity markets have historically made agricultural production unattractive to non-agrarian capital, giving family-based farms a competitive advantage (Magnan, 2012; Alston, 2004).

Discussion of the Agrarian Question was largely abandoned in the early 1990s, reflecting disillusionment with Marxist approaches following the collapse of communism, and a broader cultural turn across the social sciences. Throughout the 20th and 21st centuries, changes to technology, markets, and culture have been gradually eroding family farming's historic resilience as a form of production. The willingness of farm households to work for low or absent wages in order to maintain control of their land appears to be waning, as potential successors demonstrate their unwillingness to live in poverty (Burton and Farstad, 2020; Sutherland, 2015). Reproducing the farm (i.e. transferring it between generations) was once considered as important as reproducing the household (i.e. successfully raising children), but contemporary farming parents are similarly less willing to subject their offspring to the financial hardships of contemporary farming (Villa, 1999). Changes to technology as well as new health and safety requirements make it more difficult to socialise children into the skills and values underpinning the farming profession (Fischer and Burton, 2014). Although there is a cohort of newcomers entering the agricultural field, new entrants struggle to access the land and capital needed to establish their farms, often pursuing 'alternative' approaches both for ideological reasons, and to enable profitable production from minimum resource investment (EIP Agri, 2016). While these newcomers bring much needed innovation and energy to the agricultural sector, their numbers remain limited.

The erosion of farming culture from within is matched by broader societal critique. In the US context, Wendell (2002) wrote about how popular culture in the US aims to denigrate farmers of all kinds. What this narrative of farming does is to:

> obliterate any concept of farming as an ancient, useful, honorable vocation, requiring admirable intelligence and skill, a complex local culture, great patience and endurance, and moral responsibilities of the gravest kind.
>
> *(no page)*

Very often, discourses on climate change, biodiversity preservation, and water security identify farmers as the problem, rather than the solution. As a counterpoint, farmers can be also glorified as guardians of the 'real moral values', in the tradition of classical agrarianism (e.g. Flinn and Johnson, 1974), which often endorses "'natural'/harmonious rural-based small-scale economic activity (peasant family farming, handicrafts) and culture (religious/ethnic/national/regional/village/family identities derived from Nature)" as opposed to a supposedly more and more depraved urban society (Brass, 2000, p. 272). In a context of rising right-wing populism in the European countryside, family farming and rural life might become a contested ideological field between authoritarian populism and a renewed and progressive agrarianism based on ideas of food and local sovereignty (Mamonova and Franquesa, 2020; Scoones et al., 2018). Highly idealised notions of farming are evident in agri-tourism offerings (Flanigan et al., 2014) and computer games (Sutherland, 2020c) which present sanitised versions of farming life for popular consumption. The ideological glorification of peasants' or farmers' moral values and the commodification of farming activities as leisure practices shows as little respect to the social reality of farming as the previous technological idealism.

If we think of the idea of 'the good farmer' in the context of the wider Agrarian Question, then to think of oneself and others as a good farmer could be understood as an act of resistance to a wider notion that farmers do not matter to society. Simply continuing to exist and practice family farming could be considered an act of resistance. In a recent book on the history of African American farmers using food and farming as a survival and, thus, resistance strategy, White (2018) describes how cooperatives, demonstration farms, and African American extension programs served as what she calls collective agency and community resilience (CACR). What White shows is that farming and agriculture offer a way to carve out independent lives under difficult circumstances. In many ways her argument parallels other agri-food work on the centrality of autonomy and freedom (including repeasantisation, Ploeg, 2008) to both farmers' identities, but also their practices with particular focus on their ability to secure credit, land, contracts, and all the other necessities related to the ability to continue to build a farm operation. Further, the wider agri-food literature continues to anticipate what the future of farming might look like. This includes utopian ideas that have emerged out of some of our earlier work related to the good farmer literature (Forney et al., 2018; Forney and Stock, 2014; Nelson and Stock, 2018; Stock et al., 2014; Stock and Forney, 2014).

White's (2018) theroretical framework combines individual everyday acts of resistance with ideas of community resilience drawn from James Scott (and others). For Scott (especially in *Weapons of the Weak* [1985]), peasants (including share-croppers and other marginalised agriculturalists and pastoralists) primarily seek the ability to maintain subsistence – a right to exist. Writ large, farmers in the 20th century and the Agrarian Question as a whole have seen farmers' extinction as inevitable. Thus, the wider conflict emerges between a farmer's right to his/her existence as a farmer (in some contexts, also as a person) and modernist presumptions about what the world can and should look like which often excludes farmers. Thus, to make a claim to be a good farmer is an act of resistance. The idea of resistance then does not necessarily choose what kind of farming is better or worse, but that to keep farming is an act of resistance in and of itself. Thus, to be a new farmer is to join in that resistance (Darby et al., 2019; Stock et al., 2019).

The good farmer as solid ground for building sustainable food systems?

Resistance to change appears to be central to the 'good farmer' approach. The good farmer framing is often applied in research when the introduction of new policies or practices in agriculture has been strongly opposed by farmers and, consequently, the behavioural change promoted has failed to materialise. Here the 'good farmer' is recruited because the desired behavioural change (for example, community forestry – Burton, 1998) leads to a level of community-wide resistance that extends beyond any economic rationale and, at the same time, is justified in terms of deeply held personal beliefs about what it is to be a farmer. In these studies, the 'good farmer' is often seen as representing the influence of traditional values and, in this sense, the term can hold negative connotations as a restriction on socially desired or technological progress in agriculture. However, as we outlined in Chapter 3, this is not the case. The good farmer is constantly adapting to changing technologies, broader expectations for social behaviour (e.g. gender roles), and changing expectations for the role of agriculture in society (e.g. environmental). The good farmer could indeed be a farmer reluctant to introduce change – if the peer group happens to consist of similarly oriented farmers – but is more commonly a farmer at the top of his/her game in terms of farming and that means constantly responding to changing social, environmental, and economic conditions to stay in farming.

We see the value of the 'good farmer' concept to policy makers as not so much helping to make errant farmers adapt to policy, but rather helping policy makers understand how policies can be designed to be culturally sustainable. An example of this lies in the failure of many action-based agri-environmental schemes to promote a cultural change in attitudes to the environment. Burton and Paragahawewa (2011) argue that conventional agri-environmental schemes fail because prescribed environmental activities that are invisible in the landscape and are viewed by a farming community with no understanding of what 'good conservation farming' is, do not enable cultural and social capital to be generated in the same way as conventional agriculture. They further contend that economic incentives alone are not enough to change farming cultures because these incentives do not compensate for the loss of non-economic capital. Policies therefore need to be designed such that they:

> do more than pay unenthusiastic participants to perform tasks that are seen as unnecessary and even morally questionable, but rather are structured in a manner that enables their embedding within farming cultures to produce long-term change.
>
> *(p. 102)*

Even where change is slowed by deeply held cultural beliefs associated with the good farmer, the consequences are not necessarily negative. Rather, holding to cultural beliefs may act as a buffer against poor policy decisions. For example, UK farmers routinely point to the subsidies that funded them to remove hedgerows in the 1960s and early 1970s (a policy that was resisted by many) which were reversed shortly after joining of the European common market in 1973 as the value of hedgerows for conservation was recognised. 'Good farmers' who had not removed their hedgerows were vindicated by policies that encouraged replanting. In a profession where change is measured in generations there is a certain level of wisdom to not acting in haste – whereas for politicians anxious to be seen as fulfilling to the will of the voting population, haste is often seen as necessary. Likewise, resistance to change associated with the 'good farmer' pits years of accumulated practice-based knowledge (often geographically specific) against the scientifically based knowledge of policy makers – and introduces an understanding of locally specific agricultural conditions that are not addressed with policies implemented at the national or international scale. Moral aspects of the good farmer may further assist the development of functioning rural communities and thus help maintain the condition of rural areas and their continuation in agriculture.

In performing these roles in food production, the 'good farmer' *inadvertently* contributes to the development of sustainable food futures. However, in other cases, farming cultures are *actively* engaging in the transformation to sustainable agri-food systems. A particularly prescient example is the introduction of organic agriculture and corresponding development of a new 'organic farmer' identity with accompanying notions of what it means to be a good organic farmer (Sutherland and Darnhofer, 2012, p. 235). Here the authors witnessed the "gradual devaluation of old symbols of 'good faming'" which opened room for alternative definitions of the good farmer. Farmers could create new ways of being a good farmer without having to "convert" and change radically:

> instead of rejecting existing symbols and values outright and replacing them with others, individual farmers reflexively assess personal priorities within the broadened range of symbols and choose ways to realise them.
>
> *(p. 236)*

This produces a range of approaches from those who are committed to the organic ideals to those who convert for more pragmatic reasons – however, all make the change by attaching to elements of the good farmer identity whether the ecological benefits of organic farming or the farmer's perceived obligation to farm survival. These new identities can be built further, based on environmental elements of the 'good farmer' identity (Wheeler et al., 2018) to eventually establish a correspondence between more sustainable forms of agriculture and farming cultures that believe in the 'good' of maintaining them.

Future challenges for 'good farmers'

The economic, social, political, and environmental contexts within which agriculture takes place have been constantly changing and farmers throughout history have adapted to these changes. Today farmers face significant challenges in order to persist as farmers at the same time as addressing the multiple challenges of food security, environmental protection, and climate change. As we established in Chapters 3 and 7, symbols and norms of 'good farming' are adapted in response to changing 'rules of the game', although these responses are often delayed and somewhat indirect. In this section we reflect on the changing rules of the game and what we know – and do not know – about their intersection with the concept of the 'good farmer' in some of its possible futures. More specifically, we discuss seven evolutions of agricultural systems: the declining visibility of symbols; the increasingly contested notion of what a 'good farmer' is; the impact of advanced mechanisation and automation on farms; climate change and farming practices; synthetic proteins and food production without farmers (the 'second domestication' (RethinkX, 2019); agri-environmental schemes and payments by results; digital technologies and communication in food systems.

Declining visibility of symbols

The changing 'rules of the game' addressed in this book are primarily: the development of scientific agriculture and mechanisation, the profitability of conventional production, and the financial and ethical opportunities of organic farming. As discussed in Chapters 2 and 3, the move from implements to machinery on farms in the 18th and 19th centuries changed the symbols of good arable farming from the appearance of a verdant carpet to one of straight and parallel lines, while developments in soil science removed the necessity of evaluating soil qualities using knowledge of the wild plants that grew in and around the fields. With the increased yields that resulted and were helped, to no small extent, by the efforts of the agricultural improvers to change the definition of 'good farming', the concept of the tidy productive farm became the dominant sign of a 'good farmer'. Assisted by the financial advantages, farmers who adopted these innovations could afford to maintain their properties and invest in new machinery and were easily recognised by the condition of their livestock, fields, and buildings – displaying something to which their neighbours could aspire. The visibility of symbols of good farming is now declining – hidden behind the doors of intensive barns, concealed in agri-environmental measures and organic farming practices which yield the appearance of weedy or untended fields, and/or held as invisible knowledge of how to operate computer software or hardware. Increasingly the good farmer thus cannot be recognised through viewing fields from a distance, but must be viewed from close up and often through proxies of the skills/knowledge required for production rather than the production itself.

Contested notions of the 'good farmer'

The bifurcation of contemporary farming into businesses based on industrial production methods at one extreme, and 'repeasantisation' at the other (Ploeg, 2008) and multifunctional transitions into the provision of public goods, have yielded contested notions of what constitutes 'good farming', not just for farmers, but for society more broadly. Opportunities for direct marketing and diversification alter farming identities through integration into other fields (Brandth and Haugen, 2011). Are farmers still farmers if they work full-time off-farm or diversify into tourism, renewable energy, or other service provision which economically dwarves their farming activities? Farmers no longer solely identify as farmers, and therefore may gain their cultural or economic capital elsewhere. Sutherland (2020b) addresses part-time farming, as part of a larger analysis of the transition to hobby farming in an agrarian locale. She found that part-time farmers are not considered to be 'good farmers' by full-time farmers; the identity of 'farmer' is not accepted as multiple – a 'good farmer' 'lives and breathes farming' – a feat not possible if employed off farm for significant periods of time. Similarly, farmers who use contractors are not considered good farmers because they cannot demonstrate skilled role performance. The moniker of 'hobby farmer' is pejorative and utilised to refer to someone else's farming status.

Naylor et al. (2018) also found that hobby farmers were not identified as 'good farmers'. American evidence suggests that non-commercial farming is increasing; the US Department of Agriculture (2015) identified retirement farms (29%) and off-farm occupation farms (38.5%) as together comprising the majority of farms in the agricultural census. Although percentages of self-identified non-commercial farmers appear lower in Europe (Sutherland et al., 2019), pluriactivity has a rich tradition (Evans and Ilbery, 1993; Fuller, 1990). Opportunities for remote working or commuting make it increasingly possible to combine lucrative employment with farming practices (commercial and non-commercial), while technological advancements such as milking robots enable farmers to maintain a full time off-farm job while continuing to farm as or even more intensively than previously (Burton and Farstad, 2020). The existence of elective belonging and communities of interest (versus communities of place, as described in Chapter 7) raises the question of what is the reference point against which farmers measure their status? Where the socialisation of farmers historically has occurred within local farming communities with shared understandings of agriculture, rural areas have increasingly mixed urban/rural populations, while technological advances such as the internet and faster modes of transport make it possible for farmers to maintain a far more dispersed peer group.

Changing rules of the game also lead to reflexivity – active consideration of farming symbols and practices by farmers themselves. Recurrent food and health crises raise consumer and farmer consciousness of farming practices, challenging accepted standards or norms (Sutherland and Darnhofer, 2012). The 2013 European horsemeat scandal (where of 27 beef burger products examined, 37% tested positive for horse DNA) similarly raised consumer consciousness and critique of farming and the organisation of the supply chain. Increased regulations around traceability represent an opportunity and challenge to good farmer ideals – factory style farms are often more able to cope with the increasing regulations. As described in Chapter 7, risks associated with livestock disease have reduced the number of agricultural shows and competitions – as well as the desirability of showing livestock. Re-localisation of food through short food supply chains is increasing, but continues to serve a minority of the population, often a cohort with considerable disposable income. This raises the social justice issue of 'good farming for whom'?

Advanced mechanisation and automation

A further complication is the progressive mechanisation of agriculture. As we note in Chapter 3, machinery absorbs the skill and knowledge requirements from farm work such that, as technology advances, there are fewer and fewer ways in which the farmer is able to display good farming credentials. Globalisation of both the varieties of food produced and machinery types results in a further standardisation of symbols. The result is an impoverishment of the rich symbolic displays present in the 18th and 19th centuries. Recently the move to robotic farming in sectors such as dairy has taken the production role almost completely out of the hands of the farmers (Holloway et al., 2014a, 2014b). Bell et al. (2015) posit that there is a conjoined professionalisation and deskilling of American farmers, as advertisers portray farming as a white collar profession, while the data is analysed off-farm by 'experts'; farmers lose control of their data and are thus heavily influenced by corporate databases. Likewise, it is increasingly seed suppliers or supermarkets that determine not only what is grown on farms, but how it is grown, while increasing farm scale, corporate agriculture, and contract farming raise the question of "who actually *is* the farmer"?

The good climate farmer

A key issue for future farmers to address is the often-competing imperatives of producing increasing volumes of food and the need to address climate change. The need for the 'good climate farmer' to emerge has been recently recognised in the concept of 'climate smart agriculture' – agriculture with the multiple objectives of increasing agricultural productivity, increasing adaptive capacity to climate change, and decreasing greenhouse gas emissions and enhancing carbon sinks (Campbell et al., 2014). However, this does not necessarily lead to a single 'good climate farmer' concept. In some cases, those seeking to address the issue emphasize the productivity role, with claims that intensive production is the best way to address climate change as intensive agriculture produces lower emissions per kilogram of protein produced – even if, overall, the emissions are high. In other cases, however, the argument is made that farming needs to be less intensive as intensive agriculture is both a major user of fossil energy and releases large quantities of other greenhouse gases (in particular, methane from ruminants and nitrous oxide from nitrogenous fertiliser use) that contribute significantly to climate change. Small firms, for example, have a greater ability to supply environmentally friendly products (Krasnov et al., 2020), while Lorenz et al. (2019) found that confinement systems need to produce higher quantities of milk per cow in order to produce milk with a similar greenhouse warming potential to pasture based systems. The dual necessities of providing food and addressing climate change are thus likely to lead to many contestations of what is good farming in the coming decades.

Synthetic food and agriculture without farmers?

Of course, this assumes that farmers have any role to play in our food future. July 2019 saw a momentous event occur. Perfect Day – a company started only five years previously – won the race to become the first of 30+ start-ups to release a food product made of animal protein produced without the use of animals. Perfect Day's ice-cream was not produced by a cow, but by a yeast – a yeast that had been genetically engineered to produce six different types of cows-milk protein when fed sugar and other nutrients. The three flavours of biosynthesised dairy ice cream were made available as a limited edition release on the company's website in the United States. The public response was overwhelming. Retailing for $20 per tub (similar premium dairy ice-creams sell for $12 to $17 per tub) the product sold out within 24 hours. Reports from an invited tasting for journalists suggest that it did not disappoint. One reported that it "tasted just like real ice cream" while another observed that it may be too similar to a dairy product to appeal to the vegan market (Scipioni, 2019).

Perfect Day ice cream may be the only biosynthetic protein product to be sold to the public as yet, but it is part of a much larger movement to synthesise meat and milk that has developed since Mark Post first exhibited a burger made out of laboratory grown meat in 2013. A "proof of concept" designed to "change the discussion from 'this is never going to work' to, 'well, we actually showed that it works, but now we need to get funding and work on it'" (Fountain, 2013), the event has led to a flourishing of new start-up companies aiming to produce a variety of synthetic protein products including beef, bacon, chicken, fish, shrimp, mouse meat cat food, egg white, kangaroo, and so on. These are grown from stem cells outside of the animal – with a nutrient rich serum providing nutrients required for growth. Most meat-based start-ups predict they will have a marketable product within the coming decade, but the arrival of Perfect Day's fermentation-based ice cream in 2019 suggests that dairy farmers may be the first to be affected by these products.

What would this future mean for the good farmer? A move to the production of animal products in vitro (also termed the 'post-animal bioeconomy', Ferrari and Lösch, 2017) is the result of decades of technological development that have moved the skills of food production out of the hands of farmers and into those of corporations, engineers, computer scientists, drones, satellites, and so on.

Do we need farmers in a post-animal bioeconomy? The answer is probably yes. Synthetic meat production requires a growth serum which the industry is seeking to create from plant material, while the production of fermented products requires sugars and other nutrients to feed the yeasts. However, this would be a very different type of agriculture. To change from producing cow milk to goat milk, for example, would require simply replacing a cow-milk protein producing yeast with a goat-milk protein producing yeast – but, potentially, no other significant change in material input. Thus, rather than a physically differentiated countryside based on livestock farmers with generations of skills in livestock production and governed by notions of good farming, all that would be required for 'milk' production is sugar. With predictions that the process would involve between 77% and 91% less land use than conventional dairy (Steer, 2015), production could shrink the area required for growing sugar to only the best and flattest land – land that would be suited to further mechanisation and possibly, ultimately, complete robotisation.

In addressing this possibility Burton (2019) suggests it is unlikely that all production would shift to synthetics simply because of additional qualities that can be attached to natural protein products – such as locality, landscape, and support for rural communities. However, if bioreactors are able to churn out mass animal protein then, he argues, industrial livestock production will be outcompeted – providing the price of synthetics falls below that of industrially produced meat (which the cellular agriculture start-ups confidently predict will occur in the 2020s). In this way the post-animal bioeconomy could have significant effect on shaping the good farmer. Naturally produced protein would be sold not on the basis of nutritional content, but on production factors related to the taste, quality of animal welfare, environmental care, landscape care, and the social benefits of maintaining farming communities. In this hypothetical case the definition of the 'good farmer' – in the case of livestock at least – would be narrowed considerably from where it stands today. Good livestock farming would need to be as natural as possible in order to attract the premium prices required to remain profitable and the roles of the good farmer would adapt accordingly.

Changes to the good farmer may not be so dramatic for other types of production as there is currently no imagined way to synthesise fresh fruit and vegetables. Urban agriculture may introduce new definitions of farmer as production of fresh vegetables moves into the city – whether by using waste urban space or through the construction of purpose-built vertical farms for vegetable production. The notion of good arable farming is also unlikely to be affected in the short term. However, a recent Finnish start-up Solar Foods has developed a protein manufactured from carbon dioxide drawn from the air via a 'proprietary organism' (a bacteria that requires CO_2 and hydrogen rather than sugar) which it touts as a potential substitute for soy or wheat flour (Southey, 2019). If, as it contends, it is able to upscale production and release a commercial product in 2021 it will mean that cereal farmers and livestock farmers could face a similar threat within the next decade. Once again, the 'good cereal farmer' may need to abandon a quantitative valuation of his/her production roles in favour of emphasizing external qualities associated with the production process. With the cheap food production role taken by another sector, there may be fewer paths for the good farmer in the decades to come.

'Conservation' as a product

How else might the 'good farmer' be affected by the changing context of agriculture? One area is in the changing design of agri-environmental schemes. In particular, farmers in Europe have become accustomed to agri-environmental schemes based on a set of prescribed actions on their farms, often from a menu system – which has led to frequent selection of items on the basis of ease of management rather than ecological benefits (Butler et al., 2010). Increasingly, however, policy makers are looking at paying farmers not for performing prescribed actions but for achieving environmental results – thus addressing the perceived need for more cost-effective agri-environmental delivery mechanisms (ENRD and EC, 2010). Such schemes are gradually increasing in number across the European Union. Where this intersects with the good farmer is in that, for result oriented schemes, the outcome of 'conservation' becomes a product – measured by prescribed indicators – in the same way that conventional crops are a product from the farm. Consequently, knowledge and skills (i.e. cultural capital) in environmental management would move from being largely irrelevant (as farmers only need to follow the rules in action-based schemes) to being potentially critical to success as a farmer. In turn, and providing the guarantees of an economic return were sufficiently long-term, this would create a social value to skills and knowledge – providing peer esteem, encouraging knowledge transfer, and creating the mutual obligations necessary for the building of social capital. In this way, as the number of result-based schemes increase, environmental work could become part of the 'good farmer' identity even amongst farmers with little understanding or interest in the environment at present (see Burton and Schwarz, 2013).

Digital tools for communication and control

Another possible area where the 'good farmer' may be affected is through the development of new communication technologies and social media platforms which could lead to dramatic changes in social interaction – both within the farming community and between farmers and non-farmers. This is somewhat of a double-edged sword for the 'good farmer'. On the one hand, such technology provides an important means of communication between farmers (including, potentially, visual symbols of good farming), promoting agriculture to the public and providing a means of selling farm-based products directly to the public. However, there are also potential downsides. In a recent study, Zahl-Thanem et al. (in press) observe that in the carnivore debate in Norway social media provided internet 'trolls' with direct access to farmers who spoke out in support of predator control. The result was that some farmers felt they were unable to enter debates or put their perspective on the carnivore issue into the public arena because of the risk of trolling – examples of which include hate messaging, deliberate misreporting on predator monitoring sites, and the sending of false reports to the Food Safety Authority. This illustrates how farmers' notions of good farming can now be directly challenged by non-farming groups espousing their own views on what 'good farming' should be. The long-term implications of this are uncertain, but it is clear that the ability of mobile technologies to transfer information visually, instantaneously, and directly to specific individuals makes them potentially transformative for the transfer of symbols and notions of good farming.

The development of digital technologies in agriculture might also impact the 'good farmer' through an inflation of technologies of control, in the context of a generalisation of the audit discipline in agri-food systems (Campbell et al., 2012). Indeed, classical modes of control and governance are challenged by the extraordinary 'capacities' of Big Data surveillance (Lyon, 2014). The development of automatic data collection by connected farm tools that feed into larger centralised data bases opens farmers to both a dramatic increase in monitoring and potentially ethically questionable data extraction, accumulation, and commodification (Bronson and Knezevic, 2016; Zuboff, 2015). Information could be gathered to monitor the use of subsidies in agricultural systems (including for environmental schemes), and for better traceability and transparency in food chains. While a 'traditional' understanding of the 'good farmer' relied mostly on the role of social control by peers, the development of new technologies of accountability could give increasing weight to external auditing systems with a renewed capacity to control farmers' practices in 'real time'. Moreover, the potential extent of the influence of digital technologies in communication and monitoring highlights the need for a greater understanding of the role of technology in constructing the 'good farmer'.

Future directions for good farming research

This book has demonstrated the potential both for delving deeper into the history of the good farmer and for using various conceptualisations for advancing thinking on who is recognised as a 'good farmer' and what this means for the future of the agricultural sector. In this section we reflect on the future directions for good farming research, connecting it to advances in social thinking, research methods, and broader social and spatial justice issues.

Advancing the conceptualisation of the good farmer

In the introduction, we situated the 'good farming' literature within the cultural turn which occurred across the social sciences in the 1990s. More recently, 'relational' and 'affective turns' have occurred in social research (Clough and Halley, 2007; Leys, 2011). Although rural studies has remained largely lodged in what is now termed 'representational' research (i.e. research based on the textual and verbal expressions provided by participants), there has been a broader turn towards 'non-representational' or 'more-than-representational' approaches in the associated disciplines of geography and sociology. Non-representational thinking is an umbrella term for approaches that seek to go 'beyond': i.e. 'more-than-human', 'more-than-textual', and beyond active cognition, to consider multiple senses and the relations between human and non-human (sentient and otherwise) actants (Lorimer, 2005). Non-representational research is a broad church, bringing together actor-network theory, biological philosophy, neomaterialism, process philosophy, speculative realism, social economy, performance theory, poststructuralist feminism, critical theory, post-phenomenology, and pragmatism (Vannini, 2015, p. 3).

Non-representational thinking offers the opportunity to rethink the conceptualisation of 'good farming' in relation to actants which were previously less considered, that is, more intentionally assessing the intersecting roles of place, animals, land, and objects in co-constructing the meaning of 'good farming' and associated identities. The groundwork for the broadening of analysis is already in place – Gray (1998), for example, demonstrated how cultural norms are embedded in places, arguing that farm and farm household are 'consubstantial' – mutually constituted. Research into good farming symbols in this book and elsewhere demonstrates that these symbols have traditionally been geographically situated (developed in relation to land capacity and the types of livestock produced). Exploration of non-representational approaches appears to be a logical next step.

Non-representational thinking emphasizes the negotiated and embodied nature of social interaction. It particularly emphasizes going beyond human bodies to consider how relations are made and unmade. For example, analysis of how implementation of robotic milking systems transformed farmer understanding of the 'good dairy cow' identifies how farmer practices become refocused around 'bad cows', who perform poorly within robotic milking systems (Holloway et al., 2014a). Key in Holloway et al.'s analysis is the performance and practices of cattle themselves. Within this analysis, the agency of cattle as actors is highlighted – cows are constructed as making choices ('good' and 'bad') in how they respond to they behave within new robotic milking regimes. Burton et al. (2012) make a similar finding with respect to dairy farms in New Zealand. In this case the authors observe the creation of 'cowshed cultures' where farmer cultures, cow cultures, and the structure of the farm interact with each other to create positively or negatively reinforcing cycles of improving or worsening welfare. Both 'bad/good cows' and 'bad/good stockmen' can transfer this culture onto new farms. As described in Chapter 7, good farming symbols become literally embodied in livestock, but livestock are not passive participants in these processes, as would be suggested in most 'good farming' approaches. This line of thinking opens up avenues for considering the negotiated and embodied nature of good farming symbols within the farm setting, and indeed within the farm household, as multiple family members negotiate identity in relation to a broad set of actants.

Non-representational approaches thus draw particular attention to practices and performances: how experiences are staged, encountered, and enacted. Huttunen and Oosterveer (2017), for example, integrated practice theory with good farming concepts in their work on farmer fertilisation practices, demonstrating how the recombination of 'elements' leads to change in practices (Shove et al., 2012). Advances in technology are constructed as more than simply new 'rules of the game' to which farmers must respond, but as actants which have specific properties and attributes which lead to performances of practices which reinforce or introduce new meanings – in Huttunen and Oosterveer's case, technological advancement of fertilisation practices led to no change in the traditional meaning (of maintaining soil quality and increasing fertility) until the potential differences in meaning (e.g. environmental protection) were reflexively considered by farmers. The importance of reflexivity to changing standards of good farming was discussed in Chapter 7 (also in Sutherland et al., 2012; Sutherland, 2013).

Related to non-representational approaches, assemblage thinking is another theoretical domain with potential for developing a renewed theorisation of the 'good farmer'. Inspired by the work of Deleuze and Guattari (1988), assemblage thinking has become influential in several disciplines, notably geography (Allen, 2011; Anderson and McFarlane, 2011; McFarlane and Anderson, 2011) and anthropology (Li, 2007; Ong and Collier, 2005; Tsing, 2015). In the context of agricultural and rural studies, there is an emerging assemblage scholarship, expanding its interests to non-human actants, and paying more attention to how complex sets of relations are constituting the world. As an example, the 'Biological Economies' group (Lewis et al., 2013; Le Heron et al., 2016), gathering agri-food scholars in New Zealand, proposes a new approach to regional development that builds on a dialog between "'assemblage thinking' and 'enactive research'" (Lewis et al., 2013, p. 180). They describe the collaboration between the Biological Economies research team and several 'experimentations' developing in rural New Zealand economies, with a specific focus on the 'the making of possibilities' and a reflection on the performativity – or 'enactiveness' – of research in the process. Assemblage thinking offers also a rejuvenating theoretical framing to rethink 'rural places' in a globalising world, by focusing notably on the processes of territorialisation and deterritorialisation – understood as process of stabilisation and destabilisation – that characterise assemblages (Jones et al., 2019; Woods, 2015). As Woods (2015) suggests, considering rural places as changing assemblages of both 'material' and 'expressive' components – in De Landa's terms (De Landa, 2006) – allows understanding of the current restructuration of the rural "not as the erosion of place-difference, but as the re-assembling of places as the substitution of material and expressive components is accompanied by processes of reterritorialisation and recoding" (Woods 2015: 34–35). By articulating the materiality of the social and its expressive dimension (with includes 'representations'), assemblage thinking offers an interesting framing, we contend, to bridging the existing representational approaches of the 'good farmer' with novel, non-representational, relational and more-than-human, theorisations.

Following a similar inspiration, Forney et al. (2018) developed a preliminary framing of agri-environmental governance as assemblage, where governance is understood as emerging from repeated interaction between diverse actors – human and non-human – that are all central to the agency of the assemblage. From an assemblage perspective, farming cultures – and thereby interpretations of the 'good farmer' – can be conceived as co-constructed through the many interactions of farmers with multiple other elements of assemblages they are part of – colleagues, soil, cows, grass, policies, veterinaries, prices on the market, salesman, standards, etc. Such a framing calls for new ways of understanding cultural changes and resistances in agriculture, by widening the scope and moving the focus from farmers and farm to the wider set of relations in which they are involved. We have already mentioned the potential of approaches that would look more closely at the embodied social interactions between human and non-human actors on the farm. Another direction would be to look for 'echoes' or 'resonances' of the 'good farmer' in other places (material and symbolic) within the agri-food system. Jones et al. (2019) apply an 'assemblage methodology' to the study of the case of Welsh wool production within a globalised wool market. Their focus is not on the 'good farmer' or farming culture, however; they show clearly how localised changes have resonances across the whole assemblage. We suggest that a similar attention paid to how changes happening along the diverse assemblages in which farmers are involved – value chains, policy systems, knowledge systems, etc – might enrich our understanding of what makes farming culture transform and change, beyond the local interactions with peers. Such a research endeavour would arguably help articulate more precisely at the theoretical and empirical levels the cultural approaches of localised interpretations of the 'good farmer' with more structural analyses of agri-food system.

Advancing research methods

Non-representational and associated practice-based approaches can advance the good farming literature through more active consideration of the practices and performances of good farming identity, both at the farm level and within wider assemblages. To achieve this requires new research methods. In this section we consider opportunities for advancing the methodological approaches to understanding 'good farming'.

To date, empirical research into the 'good farmer' has largely relied on traditional, often qualitative research methods: qualitative interviewing, with some participant-observation and focus group–based research. Simply implementing visual research methods could be a step forward: visual research methods have now become relatively commonplace in sociology and geography, recognised for their utility in generating rich textual data (Sutherland, 2020a). Visual methods are surprisingly absent from much of the rural studies literature, including that addressing the 'good farmer'. Farmers clearly assess each other's performance visually – primarily through 'roadside' farming but also through evaluation of livestock quality and health at auctions and shows, etc. Farmers may also observe practices – farming skills applied in situ (e.g. cultivating fields, showing cattle at sales, etc). The symbols of good farming are thus largely visual symbols – for example tidy farmsteads, even fields of crops, healthy livestock – and form the farmer's aesthetic preference for agricultural landscapes (Burton, 2012). The importance of sensory perception more broadly could also be considered – the sounds and sensations of farming that are embedded in identities and socialisation processes. For example, the experience of interacting with livestock is not limited to visual observation – farmers touch their animals, to move them, clean them, and wash them. The unconscious way these experiences influence farmers brings us back to non-representational approaches.

Non-representational approaches emphasize that much of human response is subconscious – as Thrift (2008, p. 36), argues: "conscious awareness is fragmented and volatile" – people are not capable of actively processing the wealth of interactions that occur on a daily basis. Bourdieusian concepts underpinning much of the 'good farmer' literature are consistent with this tenet – Bourdieu clearly defines the habitus as an subconscious disposition to act; cultural capital symbols are acquired and enforced through a largely subconscious socialisation process (Bourdieu, 1984). Burton's (2004) seminal paper on the good farmer was, in part, a reaction against the then overuse of 'attitudinal studies' (i.e. the behavioural approach) in the farming literature – with researchers placing too much emphasis on reasoned behaviour (in the manner of economic studies). Perhaps it is time to take a further step away from regarding farming as driven predominantly by reasoned action. The methodological challenge we are faced with, however, is how to access and assess these subconscious processes. To date, methodological development has been far outstripped by theoretical conceptualisation within non-representational approaches, but there are some promising developments occurring which could be relevant to advancing the good farming literature.

Future work could de-centralise farmers, and instead focus on specific events, and the associated backgrounds, practices, and relations which lead to physiological responses (which reinforce or challenge the associated practices). Spinney (2015) for example, puts forth a method for combining GPS tracking and biosensing to assess response during events, an approach he argues yields more detailed verbal accounts. Looking more closely at how culturally significant behaviours are publicly and privately performed – i.e. the specific practices which encode or undermine symbolic development – could yield an improved understanding of how symbols develop and function. Examples range from the social costs of losing agricultural interaction venues – like auction marts and shows – and the overall erosion of farming identities through diversification and part-time farming and the 'making private' of skilled role performance (by increasingly indoor and desk-based performances). The role of non-farmers in shaping farming culture is increasingly evident in the fragmentation of good farming ideals and government support to achieve 'multifunctional' farming objectives. The interaction of policies and policy makers as actants in good farming practices is an important area for future research.

Non-representational approaches also draw attention to the spaces of data collection. How study participants report in one place may be very different – interviews are almost always conducted in participants' homes (i.e. on the farm) – what difference would it make to undertake these at an auction mart, a school, or a place of off-farm employment? Pilgeram (2007), for example, looks at how gender is performed at auction marts; public performance of good farming identity can also be expected to include practices of bidding, showing, dressing, and relating to livestock, facilities, and human actors. There is also a burgeoning research on urban agriculture transitions, and farm experiences as therapy for non-farmers.

The 'good farmer' literature has largely focused on farm practices and strangely enough, rarely engaged the product of farming – food. This has been consistent with a definition of farming as the provisioning activity of anonymous raw material to the food industry, very typical of a productivist regime (Wilson, 2001). However, there are niches were quality food products have contributed to creating new forms of farming pride, generally in relation to relatively short and relocalised food chains. For example, Forney and Häberli (2016) showed how different enactments of the 'local' in the Swiss dairy sector were related to contrasting identifications. In this kind of configuration, being a 'good farmer' or not is not a matter for the farming community alone to judge, but depends strongly on evaluation by other actors, mainly consumers, as well as new institutions promoting food 'quality'. Understanding how symbols of the good farmer are co-constructed in these networks or assemblages requires application of some sort of multi-sited methodologies, based on the principle of 'follow-the-thing' (Marcus, 1995) from one place to another, developing interviews and or observation with multiple types of actors (e.g. see Darnhofer, 2020). Those methods are already well established but have not been applied to the 'good farmer' to date. They also invite more novel combinations in the context of food, in relation to its material and sensuous dimensions, for example by relating the materiality of a food product to the specific interpretations of the 'good farmer' that seem to be associated with its production.

Digitisation has opened up previously unknown spaces or 'backgrounds' for farming practice. Ongoing research by Sutherland (2020c) is considering the role of on-line spaces in developing and reinforcing standards of 'good farming'. Contemporary farming computer games engage millions of players in farming practices, from feeding livestock to cultivating fields and selling commodities. These practices have the potential to reinforce or challenge player understandings of the nature of farming practices and to develop their own symbols and standards of farming. If these are combined with communication technologies and social media platforms that give the public the ability to try to influence definitions of good farming (as mentioned previously with respect to carnivores), layperson understandings of agriculture formed in gaming could have a direct influence on the definition of good farming. Game-based methods may be useful for eliciting responses to staged encounters; up to a quarter of Farming Simulator's participants (8 million copies sold to date) have direct experience of farming (Lane, 2018), suggesting that current and former farmers seek to practice farming in digital settings, implying a persistence and transmission of cultural capital outside of traditional venues. Social media similarly engages both farmers and the broader public in imagining and challenging the contemporary meanings and practices of farming.

Farming practices are also highly affective. Affect is conceptualised as a "set of flows moving through bodies of humans and other beings" (Thrift, 2008, p. 236), i.e. it is not limited to emotion or human portrayal. In simple terms, emotions are an expression of affect, but affect can extend to atmospheres and pass between people (e.g. at sporting events). More in-depth consideration of affect should yield data on how emotional responses and crowd- and family-based experiences influence behaviours, which are then latterly justified on the basis of retroactive cognitive appraisal. Applications range from analysis of the affective encounters farmers have with their livestock to assessing how farming symbols and power relations are reinforced at farmer organisation meetings. Deitz et al. (2018) have advanced participatory mapping to utilising mobile phones and GPS to develop 'emotional maps' of locales (i.e. the geophysical locations where affect is experienced and made sense of). This could usefully be applied to agricultural shows and auctions or utilised on farm tours.

Cross-cultural developments

This book reflects the focus of the 'good farmer' literature on northern/western Europe, the United Kingdom (and Commonwealth countries like Canada, New

Zealand, and Australia), and the United States. What analogues exist elsewhere and how can we build the necessary research bridges across institutional, linguistic, financial, trade, and knowledge barriers? How deeply do meanings attached to the good farmer reflect ethnic or national differences? Particularly problematic is that much of what has come to be included in the praise accorded to those deemed a 'good farmer' is based solidly on the foundations of social or colonial exclusions, the destruction of many ways of knowing, the radical flattening and erasure of nuance, difference, and contextual existence. It is in this space that we see the idea of the good farmer as one deeply connected to the various other emancipatory projects including Gibson-Graham's (2008) community economics, an emphasis on multiplicity and process in the ideas of food utopias that helps to critically examine the evolving future of food systems and the lives of those involved, biological economics, ontological turns, and wrestling with the appropriate role of technology in our fields and lives.

One of the premises of the current volume is that one person's good farmer is another person's bad farmer or not a farmer at all. These kinds of conflicts are deeply held and felt and need to be examined – these moral orders (as described by Farrell, 2015) help explain both examples of shared success (e.g. Walker, 2018, on community cooperation in Oregon; Wynne-Jones, 2017 on farmer cooperation in Wales) as well as protracted conflicts over land use decision making, technological uptake, etc. These fundamentally value-oriented conflicts also help take us away from dualistic distinctions between top-down or bottom-up or even the individual versus institutional ways of trying to understand conflict.

In line with consideration of the geographical 'other', there has been limited work specifically addressing women and gender or race within the good farming literature. Taking this further, an intersectionality framework illustrates the combinatorial effects of race, class, gender, and political geographic location that the good farming literature hitherto has missed, focusing instead on primarily heterosexual, white farmers in the United States, United Kingdom, Switzerland, and New Zealand. As argued in Chapter 6, the good farmer literature – by conforming to standard qualitative research practices of interviewing one person per household, the 'primary farmer', and not critically reflecting on the role of gender in identity formation – has contributed to the invisibility of women on contemporary farms. There is an opportunity to redress this imbalance by undertaking more focused research into women's roles utilising good farming constructs. Particular issues include how newcomers develop a 'good farmer' identity – female new entrants and women who married into farming, gender in transitions to part-time or diversified farming, and the role of women in on-farm diversification (particularly into people-facing diversification activities such as direct marketing and agri-tourism). How women shape the public image of farming is also an important issue. Women are more likely to be involved in agri-tourism activities, which actively stage farming encounters for the public and shape public perceptions. Women may thus be more influential in determining which farming practices are made visible (or invisible) to non-farming audiences.

One of the barriers to expanding our understanding of who is a good farmer in other places is language as well as structural barriers for academics in the global South. Colleagues across the world are working on what being a farmer means to be a person and a community (e.g., Dwiartama et al., 2017; Mwaura, 2017). We hope that this book, with its limitations, can help spur connections, conversations, dialogue, and shared research experience across these divides. In the end, this book and the literature it pushes forward offer a chance, a bridge, if you will, to expand not just the good farmer literature in agri-food worlds, but as a recognition that we need to be better about seeking out hidden and oppressed voices of those doing the farming, expanding and helping to expand who are recognised as farmers.

Our personal request . . .

This book represents our collective knowledge and experiences of the 'good farmer' concept built up over decades of research. While we have outlined our views, summarised existing knowledge, and even developed new knowledge related to the 'good farmer', we are clearly limited by our own limited understandings of the diverse cultural world of the farmer in a developed world context. We could have invited literally thousands of researchers, policy makers, industry representatives, and farmers themselves from across the world to join us as co-authors on the book – and each would have brought a unique perspective on what it is to be a 'good farmer'. Each would have strengthened the book. We say this to reinforce our contention that what we present here is not the final word on the 'good farmer', but rather an invitation for many others with an interest in farming culture to join us in our endeavour to understand the role of the 'good farmer' in the field and in the food system. What we desire is diversity. The development of a single concept of the 'good farmer' with defined roles that can be applied to all farmers in all situations – the optimised 'good farmer' – is not one we would welcome. It is important to remember that while the literature suggests farmers are subject to (often strong) peer pressure, this does not in any sense mean farmers all aim to be the same. If our studies have taught us anything it is that farmers seek to be innovative – to experiment with new approaches, to be better and different, to enjoy success, to conquer failure, but above all, to do so within a community where their work is understood and valued.

With the multiple moral imperatives for today's agriculture – boosting the sustainable use of resources, enhancing the natural environment, feeding the world, and so on – it is easy to believe that the objective is to 'develop' a 'good farmer' who is able to satisfy all of society's needs as well as his/her own goals. However, this is the trap fallen into by the agricultural improvers of previous centuries. There is no question that agriculture has needed to be modernised over the centuries to meet the demand for increased production and food security. However, this has often been done by drawing a simple dichotomy between 'good farmers' and 'bad farmers' with 'bad farmers' generally those following customary practices or acting under constraints improvers did not understand. This has resulted in the loss of a wide range of farming practices that today might be considered 'good farming' (e.g. the 'traditional ecological knowledge' of which ecologists frequently bemoan the loss) and serves as a warning both against trying to define a single 'good farmer' and using the concept of the 'bad farmer' to forward our own agendas.

There is indeed such a thing as 'bad farming'. As with people in any occupation, farmers can be inattentive, incapable, or incompetent – leading to poor displays of the symbols of 'good farming' and earning a reputation as a 'bad farmer'. However, when it comes to strategic choices on farm management or making the best of what is available, the existence of 'bad farming' is more questionable. As we have illustrated throughout the book, farmers are constantly having to adapt to changes in their environment such that practices that might be considered 'bad farming' at one time or in one context may be considered 'good farming' in another – with the 'bad farmer' of yesterday potentially becoming the 'good farmer' of today. For the maintenance of societal needs, it is important that we develop and hold knowledge of the values of different practices – rather than classifying them as 'good' or 'bad'. This does not mean that approaches drawing on the 'good farmer' concept should forgo critical statements. Cultivating diversity in the social and cultural dimension of farming means also promoting reflexive dialogue and debates, as well as critical engagements with how agriculture is done. The 'good farmer' concept must reflect this diversity, also to promote reflexivity in the farming worlds.

We wish to conclude by paying tribute to the subjects of our book. There are many 'good farmers' in the world, but all of the farmers we have spoken too in the many years of our collective experience are 'good people' – even though their political or environmental views are sometimes divergent from our own. We say this because they have allowed us the privilege of glimpsing into their culture, businesses, family lives, and often their homes, taking the time to explain their perspectives and trusting us with their stories. We are very aware of and thankful for this honour. As we sit as around their Formica or oak tables, lean against old farm machinery with them, or walk side-by-side through the fields, we must remember that we are observers in their world – invited guests – and it is their stories we are telling.

References

Allen, J. (2011). Powerful assemblages? *Area, 43*(2), 154–157.

Alston, M. (2004). Who is down on the farm? Social aspects of Australian agriculture in the 21st century. *Agriculture and Human Values, 21*(1), 37–46.

Anderson, B., and McFarlane, C. (2011). Assemblage and geography. *Area, 43*(2), 124–127.

Bell, S. E., Hullinger, A., and Brislen, L. (2015). Manipulated masculinities: Agribusiness, deskilling, and the rise of the businessman-farmer in the United States. *Rural Sociology, 80*(3), 285–313.

Berry, W. (2002). The prejudice against country people. *The Progressive,* April 15. www.iatp.org/news/the-prejudice-against-country-people (accessed 16 March 2020).

Bourdieu, P. (1984). *Distinction: A social critique of the judgment of taste.* Cambridge: Harvard University Press.

Brandth, B., and Haugen, M. S. (2011). Farm diversification into tourism – Implications for social identity? *Journal of Rural Studies*, *27*(1), 35–44.

Brass, T. (2000). *Peasants, populism, and postmodernism. The return of the Agrarian myth.* London and Portland, OR: Franck Cass Publishers.

Bronson, K., and Knezevic, I. (2016). Big Data in food and agriculture. *Big Data and Society*, *3*(1).

Burton, R. J. F. (1998). *The role of farmer self-identity in agricultural decision making in the Marston Vale Community Forest.* Unpublished PhD Thesis. DeMontfort University, Leicester.

Burton, R. J. F. (2004). Seeing through the 'good farmer's' eyes: Toward developing an understanding of the social symbolic value of 'productivist' behaviour. *Sociologia Ruralis*, *44*(2), 195–215.

Burton, R. J. F. (2012). Understanding farmers' aesthetic preference for tidy agricultural landscapes: A Bourdieusian perspective. *Landscape Research*, *37*(1), 51–71.

Burton, R. J. F. (2019). The potential impact of synthetic animal protein on livestock production: The new "war against agriculture"? *Journal of Rural Studies*, *68*, 33–45.

Burton, R. J. F., and Farstad, M. (2020). Cultural "lock-in" and mitigating greenhouse gas emissions: The case of dairy/beef farmers in Norway. *Sociologia Ruralis*, 60(1), 20–39.

Burton, R. J. F., and Paragahawewa, U. H. (2011). Creating culturally sustainable agri-environmental schemes. *Journal of Rural Studies*, *27*(1), 95–104.

Burton, R. J. F., Peoples, S., and Cooper, M. H. (2012). Building 'cowshed cultures': A cultural perspective on the promotion of stockmanship and animal welfare on dairy farms. *Journal of Rural Studies*, *28*(2), 174–187.

Burton, R. J. F., and Schwarz, G. (2013). Result-oriented agri-environmental schemes in Europe and their potential for promoting behavioural change. *Land Use Policy*, *30*(1), 628–641.

Butler, S. J., Boccaccio, L., Gregory, R. D., Vorisek, P., and Norris, K. (2010). Quantifying the impact of land-use change to European farmland bird populations. *Agriculture, Ecosystems and Environment*, *137*, 348–357.

Campbell, B. M., Thornton, P., Zougmore, R., Asten, P. van, and Lipper, L. (2014) Sustainable intensification: What is its role in climate smart agriculture? *Current Opinion in Environmental Sustainability*, *8*, 39–43.

Campbell, H., Rosin, C. J., Hunt, L. M., and Fairweather, J. R. (2012). The social practice of sustainable agriculture under audit discipline: Initial insights from the ARGOS project in New Zealand. *Journal of Rural Studies*, *28*(1), 129–141.

Chayanov, A. (1927). *Theory of peasant co-operatives.* Trans. David Wedgwood Benn. Columbus: Ohio State University Press.

Clough, P. T., and Halley, J. (2007). *The affective turn. Theorizing the social.* Durnham, NC: Duke University Press.

Darby, D. B., Hossler, T., and Stock, P. (2019). *New farmers 2014/2018.* Lawrence, KS: P & T Committee.

Darnhofer, I. (2020). Farming from a process-relational perspective: Making openings for change visible. *Sociologia Ruralis, n/a*(n/a).

De Landa, M. (2006). *A new philosophy of society: Assemblage theory and social complexity.* London: Bloomsbury.

Deitz, M., Notley, T., Catanzaro, M., Third, A., and Sandbach, K. (2018). Emotion mapping: Using participatory media to support young people's participation in urban design. *Emotion, Space and Society*, *28*, 9–17.

Deleuze, G., and Guatarri, F. (1988). *A thousand plateaus: Capitalism and schizophrenia.* London: Athlone.

Dwiartama, A., Tresnadi, C., Furqon, A., and Pratama, M. F. (2017). From initiative to movement: The growth and evolution of local food networks in Bandung, Indonesia. *Asian Journal of Social Science Studies*, *2*(4), 91–98.

EIP Agri. (2016). EIP-Agri Focus Group: New entrants into farming: Lessons to foster innovation and entrepreneurship. Final Report. In *EIP-Agri Focus Group: New Entrants into farming: Lessons to foster innovation and entrepreneurship*. European Commission, Brussels.

European Network for Rural Development (ENRD) and European Commission (EC). (2010). Report on the contribution of the European Network for Rural Development to the Public debate on the Common Agricultural Policy after 2013 (13/07/2010). European Commission, Brussels. http://enrd.ec.europa.eu/app templates/filedownload.cfm?id=DAB81B97-9E9B-F50F-6F18-C76EBF6B1A4A

Evans, N., and Ilbery, B. W. (1993). The pluriactivity, part-time farming, and farm diversification debate. *Environment and Planning A*, *25*, 945–959.

Farrell, J. (2015). *Battle for yellowstone: Morality and the sacred roots of environmental conflict*. Princeton and Oxford: Princeton University Press.

Ferrari, A., and Lösch, A. (2017). How smart grid meets in vitro meat: On visions as socio-epistemic practices. *Nanoethics*, 11, 75–91.

Fischer, H., and Burton, R. J. F. (2014). Understanding farm succession as socially constructed endogenous cycles. *Sociologia Ruralis*, *54*(4), 417–438.

Flanigan, S., Blackstock, K., and Hunter, C. (2014). Agritourism from the perspective of providers and visitors: A typology-based study. *Tourism Management*, *40*, 394–405.

Flinn, W. L., and Johnson, D. E. (1974). Agrarianism among Wisconsin farmers. *Rural Sociology*, *39*, 187–204.

Forney, J., and Häberli, I. (2016). Introducing 'Seeds of Change' into the food system? Localisation strategies in the Swiss dairy industry. *Sociologia Ruralis*, *56*(2), 135–156.

Forney, J., Rosin, C., and Campbell, H. (Eds.). (2018). *Agri-environmental governance as an assemblage. Multiplicity, power, and transformation*. Abingdon, UK and New York: Routledge.

Forney, J., and Stock, P. V. (2014). Conversion of family farms and resilience in Southland, New Zealand. *International Journal of Sociology of Agriculture and Food*, *21*(1), 7–29.

Fountain, H. (2013, May 12). Building a $325,000 burger. *The New York Times*. https://www.nytimes.com/2013/05/14/science/engineering-the-325000-in-vitro-burger.html

Friedmann, H. (1978). World market, state, and family farm: Social bases of household production in the era of wage labor. *Comparative Studies in Society and History*, *20*(4), 545–586.

Friedmann, H. (1980). Household production and the national economy: Concepts for the analysis of Agrarian formations. *The Journal of Peasant Studies*, 7(2), 158–184.

Fuller, A. M. (1990). From part-time farming to pluriactivity: A decade of change in Rural Europe. *Journal of Rural Studies*, *6*(4), 361–373.

Gibson-Graham, J. K. (2008). Diverse economies: Performative practices for 'other worlds'. *Progress in Human Geography*, *32*(5), 613–632.

Gray, J. (1998). Family farms in the Scottish borders: A practical definition by hill sheep farmers. *Journal of Rural Studies*, *14*(3), 341–356.

Holloway, L., Bear, C., and Wilkinson, K. (2014a). Re-capturing bovine life: Robot-cow relationships, freedom and control in dairy farming. *Journal of Rural Studies*, *33*(1), 131–140.

Holloway, L., Bear, C., and Wilkinson, K. (2014b). Robotic milking technologies and renegotiating situated ethical relationships on UK dairy farms. *Agriculture and Human Values*, *31*(2), 185–199.

Huttunen, S., and Oosterveer, P. (2017). Transition to sustainable fertilisation in agriculture, a practices approach. *Sociologia Ruralis, 57*(2), 191–210.

Jones, L., Heley, J., and Woods, M. (2019). Unravelling the global wool assemblage: Researching place and production networks in the global countryside. *Sociologia Ruralis, 59*(1), 137–158.

Kautsky, K. (1988). *The Agrarian question: In two volumes*. London: Zwan Publications.

Krasnov, E. V., Barinova, G. M., Gaeva, D. V., and Gaev, T. V. (2020). Some ways of environmentally sustainable agriculture production in the context of global market and natural barriers. In W. L. Filho, P. R. Borges de Brito, and F. Frankenberger (Eds), *International business, trade and institutional sustainability* (pp. 369–383). Switzerland: Springer.

Lane, R. (2018). Meet the real-life farmers who play farming simulator. *The Guardian*. www.theguardian.com/games/2018/jul/24/meet-the-real-life-farmers-who-play-

Le Heron, R., Campbell, H., Lewis, N., and Carolan, M. S. (Eds.). (2016). *Biological economies: Experimentation and the politics of agrifood frontiers*. Abingdon, UK and New York: Routledge.

Lewis, N., Le Heron, R., Campbell, H., Henry, M., Le Heron, E., Pawson, E., Perkins, H., Roche, M., and Rosin, C. (2013). Assembling biological economies: Region-shaping initiatives in making and retaining value. *New Zealand Geographer, 69*(3), 180–196.

Leys, R. (2011). The turn to affect: A critique. *Critical Inquiry, 37*(3), 434–472.

Li, T. M. (2007). Practices of assemblage and community forest management. *Economy and Society, 36*(2), 263–293.

Lorenz, H., Reinsch, T., Hess, S., and Taube, F. (2019). Is low-input dairy farming more climate friendly? A meta-analysis of the carbon footprints of different production systems. *Journal of Cleaner Production, 211*, 161–170.

Lorimer, H. (2005). Cultural geography: The busyness of being 'more-than-representational'. *Progress in Human Geography, 29*(1), 83–94.

Lyon, D. (2014). Surveillance, snowden, and big data: Capacities, consequences, critique. *Big Data and Society, 1*(2).

Magnan, A. (2012). New avenues of farm corporatization in the prairie grains sector: Farm family entrepreneurs and the case of One Earth Farms. *Agriculture and Human Values, 29*(2), 161–175.

Mamonova, N., and Franquesa, J. (2020). Populism, neoliberalism and Agrarian movements in Europe. Understanding rural support for right-wing politics and looking for progressive solutions. *Sociologia Ruralis, n/a*(n/a).

Marcus, G. E. (1995). Ethnography in/of the world system: The emergence of multi-sited ethnography. *Annual Review of Anthropology, 24*(1), 95–117.

McFarlane, C., and Anderson, B. (2011). Thinking with assemblage. *Area, 43*(2), 162–164.

Mwaura, G. M. (2017). Just farming? Neoliberal subjectivities and agricultural livelihoods among educated youth in Kenya. *Development and Change, 48*(6), 1310–1335.

Naylor, R., Hamilton-Webb, A., Little, R., and Maye, D. (2018). The 'Good Farmer': Farmer identities and the control of exotic livestock disease in England. *Sociologia Ruralis, 58*(1), 3–19.

Nelson, J., and Stock, P. (2018). Repeasantisation in The United States. *Sociologia Ruralis, 58*(1), 83–103.

Newby, H. (1987). Emergent issues in theories of Agrarian development. Pluriactivity and rural development. *Arkleton Trust Occasional Paper Series 2*. The Arkleton Trust Ltd., Oxford. https://arkletontrust.co.uk/wp-content/uploads/2017/12/Emergent-Issues-in-Theories-of-Agrarian-Development.pdf (accessed 16 March 2020).

Ong, A., and Collier, S. J. (Eds.). (2005). *Global assemblages. Technology, politics ans ethics as anthropological problems*. Malden, Oxford and Victoria: Blackwell Publishing.

Pilgeram, R. (2007). 'Ass-kicking' women: Doing and undoing gender in a US livestock auction. *Gender, Work & Organization, 14*(6), 572–595.

Ploeg, J. D. van der. (2008). *The new peasantries. Struggles for autonomy and sustainability in an era of Empire and globalisation*. London: Earthscan/Routledge.

RethinkX (2019). Rethinking food and agriculture 2020–2030. The second domestication of plants and animals, the disruption of the cow, and the collapse of industrial livestock Farming. *RethinkX*. www.rethinkx.com/food-and-agriculture#food-and-agriculture-download (accessed 16 March 2020).

Scipioni, J. (2019, July 16). This $20 ice cream is made with dairy grown in lab – and it sold out immediately. *CNBC*. https://www.cnbc.com/2019/07/16/perfect-day-foods-made-ice-cream-from-real-dairy-grown-in-lab.html

Scoones, I., Edelman, M., Borras, S. M., Hall, R., Wolford, W., and White, B. (2018). Emancipatory rural politics: Confronting authoritarian populism. *The Journal of Peasant Studies, 45*(1), 1–20.

Scott, J. C. (1985). *Weapons of the weak: Everyday forms of peasant resistance*. New Haven: Yale University.

Shove, E., Pantzar, M., and Watson, M. (2012). *The dynamics of social practice: Everyday life and how it changes*. London: Sage Publications.

Southey, F. (2019). Solar foods makes protein out of thin air: 'This is the most environmentally friendly food there is'. *Food Navigator*, July 15. www.foodnavigator.com/Article/2019/07/15/Solar-Foods-makes-protein-out-of-thin-air-This-is-the-most-environmentally-friendly-food-there-is

Spinney, J. (2015). Close encounters? Mobile methods, (post)phenomenology and affect. *Cultural Geographies, 22*(2), 231–246.

Steer, M. (2015). A comparison of land, water and energy use between conventional and yeast-derived dairy products: An initial analysis. Report prepared for perfect day. University of the West of England, Bristol.

Stock, P. V., Darby, D. B., and Hossler, T. (2019, October 29). Experiments in the field. *Visual and New Media Review, Fieldsights*. https://culanth.org/fieldsights/experiments-in-the-field

Stock, P. V., and Forney, J. (2014). Farmer autonomy and the farming self. *Journal of Rural Studies, 36*, 160–171.

Stock, P. V., Forney, J., Emery, S. B., and Wittman, H. (2014). Neoliberal natures on the farm: Farmer autonomy and cooperation in comparative perspective. *Journal of Rural Studies, 36*, 411–422.

Sutherland, L-A. (2020c). The 'Desk-Chair Countryside': Affect, authenticity and the rural idyll in a farming computer game. *Journal of Rural Studies, 78*, 350–363.

Sutherland, L-A. (2013). Can organic farmers be 'good farmers'? Adding the 'taste of necessity' to the conventionalization debate. *Agriculture and Human Values, 30*, 429–441.

Sutherland, L-A. (2015). EIP-Agri Focus Group New entrants into farming: Lessons to foster innovation and entrepreneurship. Discussion paper. In *EIP-Agri Focus Group New entrants into farming: Lessons to foster innovation and entrepreneurship*. Discussion paper, ed. EIP-Agri Agriculture and Innovation. https://ec.europa.eu/eip/agriculture/sites/agri-eip/files/fg14_new_entrants_starting_paper_2015_en-v2.pdf.

Sutherland, L.-A. (2020a). Agriculture and inequalities: Gentrification in a Scottish parish. *Journal of Rural Studies, 68*, 240–250.

Sutherland, L.-A. (2020b). Finding 'Hobby' farmers: A 'Parish Study' methodology for qualitative research. *Sociologia Ruralis, 60*(1), 129–150.

Sutherland, L.-A., Barlagne, C., and Barnes, A. P. (2019). Beyond 'Hobby Farming': Towards a typology of non-commercial farming. *Agriculture and Human Values, 36*(3), 475–493.

Sutherland, L.-A., and Darnhofer, I. (2012). Of organic farmers and 'good farmers': Changing habitus in rural England. *Journal of Rural Studies, 28*(3), 232–240.

Thrift, N. (2008). *Non-representational theory: Space, politics, affect.* Abingdon, UK and New York: Routledge.

Tsing, A. L. (2015). *The mushroom at the end of the world. on the possibility of life in capitalist ruins.* Princeton: Princeton University Press.

US Department of Agriculture (2015). *Farm typology.* Vol. 2. Subject Series. Part 10. AC12-S-10. National Agricultural Statistics Service. www.agcensus.usda.gov/Publications/2012/Online_Resources/Typology/typology13.pdf. (accessed 16 March 2020).

Vannini, P. (2015). Non-representational research methodologies: An Introduction. In P. Vannini (Ed.), *Non-representational methodologies: Re-envisioning research* (pp. 1–18). New York: Routledge.

Villa, M. (1999). Born to be farmers? Changing expectations in Norwegian farmers' life courses. *Sociologia Ruralis, 39*(3), 328–342.

Walker, P. (2018). *Sagebrush collaboration: How Harney County defeated the takeover of the Malheur National Wildlife Refuge.* Corvallis: Oregon State University Press.

Wheeler, R., Lobley, M., Winter, M., and Morris, C. (2018). "The good guys are doing it anyway": The accommodation of environmental concern among English and Welsh farmers. *Environment and Planning E: Nature and Space, 1*(4), 664–687.

White, M. M. (2018). *Freedom farmers: Agricultural resistance and the black freedom movement: Justice, power, and politics.* Chapel Hill: University of North Carolina Press.

Wilson, G. A. (2001). From productivism to post-productivism and back again? Exploring the (un)changed natural and mental landscape of European agriculture. *Transactions of the institute of British Geographers, 26*(1), 77–102.

Woods, M. (2015). Territorialisation and the assemblage of rural place: Examples from Canada and New Zealand. In J. Dessein, E. Battaglini, and L. Horlings (Eds.), *Cultural sustainability and regional development: Theories and practices of territorialisation.* Abingdon, UK and New York: Routledge. 29–42

Wynne-Jones, S. (2017). Understanding farmer co-operation: Exploring practices of social relatedness and emergent affects. *Journal of Rural Studies, 53*, 259–268.

Zahl-Thanem, A., Burton, R. J. F., Blekesaune, A., Haugen, M., and Rønningen, K. (in press). The impact of wolves on psychological distress among farmers in Norway. Accepted by *Journal of Rural Studies.*

Zuboff, S. (2015). Big other: Surveillance capitalism and the prospects of an information civilization. *Journal of Information Technology, 30*(1), 75–89.

INDEX

Note: Page numbers in *italics* indicate a figure on the corresponding page.

Made in the USA
Las Vegas, NV
19 September 2026

125ecba9-5ac5-42f7-8e9b-ba9b4166d0e4R01